Innovation in Rural Development and Social Work

ABOUT THE AUTHOR

Prof.Menon Shroff, is an Ecologist and Environmental Specialist with more than 28 years of academic and consulting experience. He holds a Ph.D in geography from the Univeristy of Birmingham UK and wrote the Ph.D and Master Academic Programs of Environmental Sciences for the Universidad de Guayana, Venezuela.

ABOUT THE BOOK

Rural social work may be defined as the practice of social work in any setting identified by those present in the area as rural. This may include farming, mining, fishing, logging, or ranching communities and small towns and villages of many types. There is ongoing debate over how to definitively categorize rural areas, but generally such debate is mostly restricted to those who believe themselves to be, or have an interest in, rural areas, however defined. The most widely used definitions involve uses of census data for population and area. Rural development and poverty alleviation programmes are implemented on a decentralised basis, keeping in view the large geographical areas, the administrative requirements and the need to involve grassroots-level officials and the community in the implementation of the programmes. At the central level the Ministry of Rural Areas & Employment has been implementing these programmes. Social work is a profession for those with a strong desire to help improve people's lives. Social workers assist people by helping them cope with and solve issues in their everyday lives, such as family and personal problems and dealing with relationships. Some social workers help clients who face a disability, life-threatening disease, social problem, such as inadequate housing, unemployment, or substance abuse. Social workers also assist families that have serious domestic conflicts, sometimes involving child or spousal abuse. This book's clear, authoritative chapters describe the declining opportunities available in rural areas including the social, agricultural, dairy development project, trends in poverty and rural poverty eradication.

Innovation in Rural Development and Social Work

Prof.Menon Shroff

AMIGA PRESS. INC
Delhi -110051

AMIGA PRESS. INC
H.O :- 28-A, Indra Park, Chander Nagar
Delhi :- 110051 (India)
Tel :- 011-22504028, +91-98736-57000
Email :- info@amigapress.co.in
amigapub@gmail.com
Website :- www.amigapress@co.in

This edition is published by Amiga Press.Inc

ISBN : 978-93-87295-19-3

First Published 2019

Printed and bound by Amiga Press Inc. New Delhi-India

Preface

Rural social work may be defined as the practice of social work in any setting identified by those present in the area as rural. This may include farming, mining, fishing, logging, or ranching communities and small towns and villages of many types. There is ongoing debate over how to definitively categorize rural areas, but generally such debate is mostly restricted to those who believe themselves to be, or have an interest in, rural areas, however defined. The most widely used definitions involve uses of census data for population and area.

The nomenclature 'Rural Sociology' indicates that it is the scientific study of Rural Society. Broadly speaking, it deals with the systematic study of Rural Society, its institutions, activities interactions, social change etc. It not only deals with the social relationship of man in rural environment but also takes urban surroundings into considerations for a comparative study. So Rural Sociology primarily deals with the description and analysis of groups of various kinds as they exist in rural environment.

Most rural societies are relatively 'integrated', in the sense that the various components of life are closely interrelated. Rural people do not easily recognize the distinctions which planners, extension workers and other government officials make between, for example, the responsibilities of different agencies or 'economic' and 'social' planning, because in their own lives all these things affect each other.

The contribution which social services make will depend to a considerable extent upon personnel and leadership. A general understanding of the philosophy and history of social work, the structure and functions of society, the nature and extent of social problems, the methods and techniques of social work, and of the details of the programmes and how best their results may be assessed, will help improve the quality and efficacy of all services organized by State and private agencies. The training of social workers should of course include knowledge of conditions prevailing in fields in which they are to work, and social workers must possess the spirit of service and the character and energy to execute programmes despite handicaps and limitations and with such resources-as may be readily available.

Sustainable economic development emphasizes the enhancing of existing local assets without degrading their quality. Political sustainability hinges upon the support of the majority of the community and must be effective over the long-term. Social sustainability requires that a project or initiative be integrated into, and connected with, a community's particular social structure. Environmental sustainability relates to the impact of a project or proposal on the local environment, with enhancement rather than degradation being the important outcome.

Social work is a profession for those with a strong desire to help improve people's lives. Social workers assist people by helping them cope with and solve issues in their everyday lives, such as family and personal problems and dealing with relationships. Some social workers help clients who face a disability, life-threatening disease, social problem, such as inadequate housing, unemployment, or substance abuse. Social workers also assist families that have serious domestic conflicts, sometimes involving child or spousal abuse.

Additionally, they may conduct research, advocate for improved services, or become involved in planning or policy development. Many social workers specialize in serving a particular population or working in a specific setting. In all settings, these workers may also be called licensed clinical social workers, if they hold the appropriate State mandated license.

This is a reference book. All the matter is just compiled and edited in nature, taken from the various sources which are in public domain.

This book's clear, authoritative chapters describe the declining opportunities available in rural areas including the social, agricultural, dairy development project, trends in poverty and rural poverty eradication.

—*Editor*

Contents

1

Rural Development

When it was established by His Highness the Aga Khan in 1984, the Aga Khan Rural Support Programme in India (AKRSP) started field operations in the state of Gujarat. Since then, AKRSP has reached over 500,000 beneficiaries in over 1100 villages in the states of Gujarat, Madhya Pradesh and Bihar. Over 4,000 village organisations have been created.

Water conservation and sustainable natural resource management are at the heart of AKRSP's work. In coastal areas of Gujarat where saline water has encroached on fresh water supplies, AKRSP has pioneered methods of remediation and community management of these resources. Its methods have been replicated in other states.

AKRSP first began working in three areas of Gujarat: the "tribal" areas in the south where some of the poorest communities live and where natural resources have been poorly managed; the coastal region and the area surrounding the Gir Forest, which now suffers from increasing groundwater salinity due to over-pumping; and Surendranagar District, which is one of the most drought-prone regions in Gujarat.

In 2004, AKRSP expanded to the neighbouring state of Madhya Pradesh, which had a high degree of food insecurity. In 2008, AKRSP was asked to contribute to rural development in Bihar. Today, AKRSP's programmes in India have four main components:

- Economic development: agricultural and non-agricultural interventions that help improve food security, increase agricultural incomes and reduce risks of the landed farmers and provide livelihood options for the poor and landless farmers.
- Social development: programmes that address social inequities and integrate everyone regardless of gender, caste or "tribal" origins in the decision-making process, providing a "voice" to the marginalized.
- Basic services: to fill the perennial need for water, energy and fodder, AKRSP works to provide infrastructure for drinking water, technical

know-how as well as infrastructure for generating alternative energy and fodder growth for cattle.

- Improved governance: AKRSP encourages the formation of a range of organisations at the village, sub-village and multi-village level that are responsive to the needs of their communities and influence local governance structures; it also works at the state level to influence government policies regarding the rights of such groups.

Improved Food Security and Increased Incomes

The major goal of AKRSP's agricultural programmes has been to ensure food security, increase net incomes and reduce risk for farmers. AKRSP achieves these goals by working to improve the quality of land, increase the availability of water for irrigation and introduce new inputs or technologies that improve agricultural productivity. Focusing on landless farmers, families with small holdings and farmers with three acres or less, it has introduced a range of options from systems to improve rice and wheat yields to small scale vegetable farming. In 2010, for example, farmers adopting intensification techniques proposed by AKRSP reported a 30 percent rise in yields. Landless farmers have planted creeper vegetables such as gourds and beans. Composting and fertiliser production have also been demonstrated. Drip and sprinkler irrigation techniques have been piloted and scaled up. In Gujarat, the rising demand for vegetables and dairy has spurred farmers to change crops; in other areas, AKRSP has helped farmers to begin growing fruit trees including papaya, sapota, mango and lemon to meet shifting demand. It has promoted village organisations to take up collective agri-input supply and marketing to ensure that poor farmers are not exploited by local traders and have access to timely and quality seeds and other inputs and are linked to market. Animal husbandry and care are also important parts of the programme.

Soil and Water Conservation

Undertaken with support from government agencies, AKRSP has operated a number of programmes dedicated to the sustainable management of soil and water, the main natural resources available to India's rural populations. Its soil and water conservation measures have improved over 40,000 hectares of land. Interventions encompass the formation of watershed groups and participatory irrigation management in Gujarat and Madhya Pradesh; dozens of irrigation and ground water recharge systems; promotion of micro-irrigation devices like drips and sprinklers; river basin management, including the construction of over 1000 check dams and irrigation tanks; and other watershed management measures.

AKRSP was selected as Project Implementation Agency for the Mahatma Gandhi National Rural Employment Guarantee Scheme (MGNREGS) in Madhya Pradesh, which ensures that employment generation schemes lead to productive assets. Distress migration has been reduced by 70-90 percent for farmers and by 30-50 percent for agricultural labourers, according to the research study carried out by the international water management institute titled "Agrarian Transformation Among Tribals: From Migrants to Farmer Irrigators". AKRSP was the first NGO to implement Participatory Irrigation Management (PIM) in Gujarat. Today, there is a PIM Act in Gujarat. Even AKRSP's watershed approach has been recognised by Government through a citation of AKRSP in the revised national watershed guidelines. Activities have included motivating communities to raise and protect common land (forest land and village commons), tree planting campaigns and watershed-related reforestation. Based on its pilots, it collaborated with other NGOs to draft a Joint Forest Management policy, which was eventually adopted by 19 states of the country and benefited thousands of forest users in the country. AKRSP has planted over 12 million trees in the past 25 years. Over the years it has also introduced a range of alternatives to biomass consumption, including biogas plants and solar lanterns. Its agricultural and farm forestry measures have covered over 11,000 hectares.

Climate Resilience

In areas impacted by drought, agriculture pests or salinity ingress due to climate change and other factors, AKRSP has worked to introduce alternative crops that are more resilient in the face of these changes. Alternative crops have also helped farmers raise incomes. In saline areas, it has introduced saline-resistant crops, such as sapota. AKRSP has also helped farmers raise incomes through other sources of incomes, including organic fertiliser production, handicrafts, bamboo furniture, honey production and other non-farm sources of income.

Alternative Energy

Even before global warming became a concern worldwide, climatic conditions in AKRSP programme areas had forced it to explore alternative energies, first through biogas projects and more recently through windmills and solar energy. Seeking a solution to the drudgery of rural women who spend two to three hours daily collecting fuel wood, AKRSP first piloted biogas plants in Gujarat. AKRSP has constructed over 10,000 household biogas units, many of these attached to household toilets. In Bihar, where the electricity supply is usually not available (despite electric lines being in place), AKRSP piloted solar lanterns that can be charged at a central charging station run by an entrepreneur. To address the destruction of the Gir Forest because of fire-wood collection, AKRSP has also piloted biogas plants, solar cookers and windmills, including a low-cost windmill for water pumping. As of 2010, AKRSP had supported the installation of nearly 14,000 biogas, solar or wind systems. The ultimate aim of the programme is to reduce the consumption of biomass and non-renewable sources such as kerosene and reduce the drudgery and indoor pollution affecting rural women.

Potable Water and Sanitation

Potable drinking water programmes have been a priority, whether in saline areas of Gujarat or in flood-prone Bihar, where bacterial contamination has been very high. In Gujarat, AKRSP partners with the government and the village panchayat to scale up interventions which provide potable water through household and village drinking water schemes. AKRSP has built or rejuvenated 125 drinking water schemes, 200 percolation wells and over 10,000 roof rainwater harvesting structures. As a result of these efforts, over 40,000 women have access to potable drinking water. Where practicable, water is piped to the home; in other locations, water harvesting systems for the home have been introduced. One of the most important impacts of these systems has been the reduction of drudgery for women and girls. The Nirmala water testing lab in Surendranagar in Gujarat, set up by AKRSP, tests water samples and trains villagers in ways of improving water quality. AKRSP has promoted a Coastal Salinity Prevention Cell which partners with the government to provide drinking water access to over 250 coastal villages. In Bihar, AKRSP tests water quality and supports low cost interventions which reduce bacterial content substantially.

Community Organisation

One of the characteristics of ultra-poor communities has been a lack of organisation. In response, AKRSP has worked to mobilise communities in its programme areas so that they can plan for and implement their own development plans. Over 4,000 village organisations have been established. With AKRSP's assistance, these organisations support the construction of infrastructure, improve productivity, create awareness about an issue (e.g., water conservation or natural resource management) and liaise with the government over development concerns. The aim of these programmes is for the committee level organisations to become entirely self-reliant, with the skills to federate and form linkages with government programmes. Women's federations in south Gujarat, for example, already have their own office buildings and access funds from the National Bank for Agricultural and Rural Development.

AKRSP (India) played a major role in influencing the Joint Forest Management (JFM) policy and was instrumental behind issuing a national-level Government Resolution on it.AKRSP (India) was the first NGO to implement Participatory Irrigation Management (PIM) in Gujarat. Today, there is a PIM Act in Gujarat. Even the watershed approach of AKRSP (India) has been recognised by Government and it has been mentioned as one of the key NGOs in the revised watershed guideline of the Central Government.

Savings and Credit

To smooth out the shocks of sudden costs, such as healthcare bills or the purchase of seeds or tools, AKRSP has supported the creation of self-managed community-based savings groups for many years. Typically, groups of 15-20 women are supported through basic financial literacy training. Savings groups have saved over Rupees 35 million. In Gujarat and Madhya Pradesh, these groups are linked to banks and access credit for their members.

Education

AKRSP also has worked on a variety of education initiatives ranging from adult literacy to computer training to early childhood development. One of its largest programmes is training rural youth in computer basics; 25,000 youth have been trained in all three states. AKRSP created 30 Community-based Technology Learning Centres. In Bihar, it supports 42 Learning Support Centres, which focus on children ages 6-11, as well as 42 early childhood development centres that train mother-teachers to impart education to 3-6 year olds.

Sharing Lessons Learnt

In a large country like India, there is a limitation to the reach of NGOs. AKRSP believes that it needs to share its learning with many more villagers and stakeholders through the following strategies:

- Training and Capacity Building: Two training centres located in rural Gujarat have trained more than 20,000 villagers, government and NGO staff so far. A dedicated training unit ensures that training quality is constantly upgraded.
- Research, Documentation and Dissemination: AKRSP undertakes research into its own work and disseminates its learning in state and national workshops. A radio programme in tribal dialect reaches out to more than 30,000 tribals in South Gujarat.

- Policy Influence through Networks and GO-NGO Collaborations: AKRSP is an active member in several networks on drinking water, agriculture, etc. and has helped influence national and state policies, as well as programmes in forestry, irrigation, drinking water and rural livelihoods.

VOLUNTARY ORGANISATIONS FOR RURAL DEVELOPMENT IN INDIA

I don not disregard the niches which market forces and state/public agencies leave unfilled. But I argue that these niches can be filled not necessarily only by the third sector or voluntary organisations but by the 'developmental deviants' or 'entrepreneurs' or 'volunteers'. These volunteers while remaining in mainstream public or market organisations, can create new alignments (by playing an insurgent function) between social needs and institutional support. The excessive attention on voluntary organisations by aid agencies – national and international – seems misplaced in so far as these agencies almost completely neglect the 'developmental volunteers' (DVs).

By reducing pressure for reform and self-renewal on the public and market agencies, the support to only NGOs may become dysfunctional not in distant future. NGOs led by managers or leader from often urban context by their own creativity suppress or fail to nurture creativity of local disadvantaged groups. The social change thus becomes slower because it becomes dependent more and more on external leaders.

In a classic paper on Organisational Forms using transaction cost framework, Ouchi (1981) suggested that markets worked best under competitive condition of information dissemination and existence of a moderate amount of trust. The clan form evolved when trust is maximised, supervision and control are minimised. The bureaucratic form signify absence of trust and thus maximisation of transport. NGOs can behave like a private sector firm, a bureaucracy or a clan with feudal or democratic values. There is nothing innate in the category of an organisation that should make it suitable for any purpose per se. It is for this reason that Sheth and Sethi (1989) face a problem in classifying NGOs.

Social purpose does not necessarily underlie a non-governmental form of the organisation. NGOs could also include institutional grouping such as the underground Mafia networks or country liquor manufacturers networks.

It is also important to note that NGOs do not emerge in certain sectors as easily as other. For instance, in the field of Science and Technology there are far fewer NGOs which are engaged in development of technology adapted to region specific conditions. Though there are many which are engaged in the process of technology transfer.

Rural development as a part of social change is defined here as a process of

(a) Widening the decision making horizon and

(b) Extending the time frame for appraising investment and consumption choices by the rural disadvantaged people collectively and not necessarily at village but even at higher levels of aggregation.

There seems to be a conceptual confusion in the political discourse on organisation of equity and design of delivery systems. It is assumed that providing equal opportunities to everybody will fulfil the constitutional obligations. What is ignored is that historical disadvantage has not been uniformly experienced by the poor in all the resource markets. Thus three groups for instance, may have varying disadvantage in terms of their respective access to resources, say, to land, labour and livestock, their skills to convert resources into investment and the quota of assurances or ration of assurances required by them to deal with the uncertainties and risks in each of the resource markets. It is not enough therefore, to provide equal opportunity to all the three groups in all the three resource markets. The challenge is to organise inequity in the delivery of services such that historical disadvantages and knowledge resources of the poor people could be adequately compensated or built upon.

NGOs have faltered in this regard as often as public institutions. The notion of profit is considered bad, the services are not charged, the clients are made thus dependent upon services for which they have not paid. And thus when the aid is withdrawn the system breaks down and people sometimes ar worse off because of their impaired potential for self-reliance.

A question can be raised as to whether one should conceptualise the role of NGOs as systems which remedy errors in the delivery of goods through markets of public systems; do they fill gaps left by both; do they signal the weak links so that systems can repair its inadequacy well in time or do they provide space for innovations that require different norms of discourse and governance than are possible in public and marketing systems. In our view different NGOs could meet different needs mentioned above. What is important is to recognise that not just the organisational form but the human urge to take a voluntary step to go beyond once formal institutional responsibility is what constitutes the biggest challenge today. The form of organisation are important in so far as the boundaries have to be drawn for use of any resource in sustainable manner. The accountability towards society and mechanisms of democratic governance would ultimately determine whether one form has an advantage over another.

Voluntarism may affect any one or more subset of developmental triangle of access, assurance and ability of the people and thus many remain restricted in its impact. Propositions given next deal with the way voluntarism has been related to the process of social change in India. Given the range of experiences it is indeed a synoptic account.

Voluntarism triggered by a natural crisis like flood, drought, cyclone etc. may legitimise entry of outsiders in a given region. But depending upon the mobilisation process NGOs emerging in response to such crisis often diversify into other areas of social development and remain community oriented rather than class oriented.

Several Church based NGOs came into existence when international aid agencies offered relief at the time of Bihar famine in Sixties. Most of the relief was in the form of consumables like foodgrains, clothes, medicines, etc. The organisational structure for distribution of such aid was different from the structure for managing durable assets like rigs for drilling wells, transport, buildings, etc. From relief to reconstruction attracted several young people. Instead of going back to pursue their professional career, they remained behind to organise people in some cases, manage food for work programme, drill wells or provide health and educational facilities in other cases.

Many aid agencies sought legitimacy through relief but indulged in other interventions subsequently. The reaction of state agencies was to incorporate such volunteers or voluntary organisations as appendage of public relief and development programmes. Such an incorporation also took place in case of many NGOs which came into picture much later. An interesting feature of these organisations was that having begun with community approach (relief was needed by all – big or small), they continued to use an electric approach to development. Social conflicts were just noted by some and participated in by others. The institutionalisation of voluntarism in intermediary support or funding organisations or grassroot organisations gave a techno-managerial start to the intervention strategies. A negative feature of such aid was that in regions prone to frequent natural calamities, people started losing their self-help initiative. State relief in the form of employment or food was not linked with mobilisation of voluntarism among people.

Dependency so created made the task of many radical NGOs even more difficult. People could not understand why mobilisation around a radical ideology should be a reason for foregoing immediate material benefits. Voluntarism triggered by man-made disasters such as Bhopal tragedy could get caught in the dilemma of legitimising state's indifference by becoming part of urgent relief and rehabilitation vis-à-vis questioning the basis of tragedy and complicity of state in its consequences.

Ravi Rajan (1988) while analysing rehabilitation and voluntarism observed four distinct styles,

(i) Intervening organisation took upon provision of relief and rehabilitation as primary task, became dependent on government and with dimunition in governments' own commitment to the cause, collapsed soon;

(ii) Volunteers served as 'conscience keepers', pursuing change through systematic research reports;

(iii) Trade union activists demanding charge of the industrial plant to provide employment through alternative use of plant and machinery; and

(iv) Perhaps the most significant strategy by volunteers was rejecting the idea of voluntarism as propounded by the state.

Sustained mobilisation, struggle for better relief, access to medical data, questioning the secretiveness on the part of the government (seen as a sign of its complicity with MNC – Carbide), legal activism and questioning the right of government to price the life of poor so low. Voluntarism of this nature is difficult to mobilise in backward rural areas given the dispersed nature of settlements and weak social articulation, low media attention and poor networking among interventionists.

Voluntarism manifested in sixties in the form of;

(a) Protest against agrarian disparities (in the form of naxalite left violent movement), and

(b) For social reconstruction (initiatives by students, professionals in the mainstream organisations, or voluntary organisations) has undergone a sea change in the wake of recent economic liberalisation.

Radical groups using violent means of social change had sought support essentially from Maoist philosophy. After the Chinese aggression in 1962, the covert support to these groups had increased. With the decline in this support these groups faced some of the similar dilemmas as faced by the other groups supported from the right. The income disparities intensified after first phase of 'green revolution'. Technological change had provided the spur for a large number of young people particularly from West Bengal and Andhra Pradesh to plunge into the field of violent social change. The attempt was to annihilate rich farmers and other symbols of perceived oppressive classes or considered class enemies. There were few groups like the one in Warli in Maharashtra which decided to remain dependent upon local people even for their subsistence needs. The possibility of such groups losing local support were remote. Likewise due to the ability of another group in Andhra Pradesh to dispense justice at low cost in cases of intra and inter-group conflicts among the tribals, even the lower level bureaucracy started using these forum for the redressal of their grievances.

Another stream of volunteers entering the field of rural development came with innovative ideas for providing relief during 1964-66 drought in different parts of the country. These volunteers became crucial instruments of social dynamics. War in 1965 with a neighbouring country had led to slowing down of U.S. aid to India. Search for indigenous alternatives became intense and legitimacy of voluntarism increased.

The 1966 to 1971-72 period was full of economic crises. Economic environment in the preceding decade and had been aimed at closure of Indian economy through import substitution. Droughts, wars (1965,1971), devaluation of currency and inflationary pressure created an environment of social unrest in organised and unorganised sector. The starvation deaths are supposed to have been eliminated (almost) after 1965-67 drought. Maharashtra started Employment Guarantee Scheme during drought of 1972. In the wake of large scale violence in 1966-67 by left radical groups, the report of a confidential enquiry committee by a committed civil servant (Appu) set up by Ministry of Home argued for immediate thrust towards target group oriented programmes of rural development suited to location/ecology and class specific needs.

The Small and Marginal Farmer and Agricultural Labourer Development Agencies, Drought Prone Area Programme, Tribal Development, Hill Area development plans etc. followed. The decentralised development in the policy was accompanied by greater political centralisation for a little while (1970-77). A movement based on Gandhian values calling for total social revolution was spearheaded by Jaya Prakash Narayan in 1973-74. It attracted a large number of young people particularly in Gujarat, Bihar and Maharashtra many of whom have continued with voluntary work.

Government declared emergency during 1975 till 1977 after prolonged railway strike and even urban people realised for the first time the implications of a non-democratic coercive state. Voluntarism, was also sought out as a sign of despicable deviance. People had option to be incorporated into the repressive state structure or be put in jail or victimised. Most chose soften options but many did not. The Post 1977 phase of change in the political continuity through single party rule brought many Gandhians committed to decentralised development into the mainstream. Tax concessions for voluntary initiatives by commercial companies were introduced first time by Janata government in 1978. Lot of innovative organisations came into being. Several developmental volunteers working in commercial organisations found this an opportunity for exploring new organisational space. Some misused this option but many did not.

For the first time, professional and young activists were offered competitive salaries in addition to autonomy for work unheard of in the mainstream organisation by and large. This was also accompanied by a change in the policy of international aid agencies which started shifting from funding better implementation of government programme bureaucratically to better implementation by NGOs. It was rather unfortunate that creative avenues in NGOs got generally fossilised due to the proximity to state and participation in implementing standardised programmes.

Change of government in 1980 and restoration of the rule by Congress party led to;

(a) Expected withdrawal of tax concessions,

(b) Centralisation of voluntarism i.e. companies could contribute to Prime Minister's fund for rural development and seek fresh grants from it for action programmes,

(c) Direct transfer of funds from a commercial balance sheet tot eh social (less easy to account) balance sheet was stopped,

(d) Standardisation of developmental programmes such as IRDP (Integrated Rural Development Programme),

(e) Withdrawal of higher allocations in IRDP for backward areas and putting them at par with the rest;

(f) Merger of earlier eco-adaptive or responsive programmes into uniform standardised IRDP with credit-linked subsidy as the dominant mode of relationship etc.

Another interesting development was the phenomena of the return of naxalite underground workers into the mainstream non-violent but articulate strategies of social change. For the first time, several ex-naxalities fought election in 1977 and some got elected also.

The social space for alternative development was filled by volunteers having varying backgrounds:

(a) Ex-radicals,

(b) Liberal or social democrats dissatisfied with the working of state and wanting to the influence distribution of resources,

(c) Enthusiastic urban activists looking for a career, failing to get one, returning to mainstream professions rather quickly,

(d) Young professionals with technical or other disciplinary background launching action-research projects or supporting other professional groups (notable being PRADAN set up by Mahajan, an alumni of IIM-a),

(e) Retired civil servants, ex-Gandhians, lawyers etc. who formed independently or with the support of aid agencies large NGOs and

(f) Quasi-state organisations promoted to provide technical, financial, marketing or other support to NGOs, artisans and other beneficiaries of state sponsored developmental programmes.

At the time when social space for volunteers was widening, opportunities for career growth in mainstream organisation also started increasing. First phase was the growth in Banking sector after nationalisation in 1969. A large number of bright young men and women with background in science, humanities or engineering joined banks, insurance corporations and other such systems. 'Brain Draft' as against 'Brain Drain' took its heavy toll by depriving academic disciplines of bright students and luring some 'professionals on the margin' away from other direct social development systems. Post 1980 boom in consumer goods industry and continued growth of banking and other public and private ventures further increased the flow of young people towards such careers. The opportunity cost of those who chose to work in NGOs did indeed increase.

The question we want to address next is, what are the processes by which voluntarism in the mainstream organisations can complement the efforts of NGOs in not merely bringing about social change at micro level but also influencing public policy in favour of disadvantaged?

Implications for Actions and Research

Generating extra-organisational space for developmental volunteers within mainstream organisations is a necessary conditions for sustainable social development.

A recent study on Bank and NGO Interface for poverty alleviation in backward regions, noted that

(a) There was no NGO working in 50 most backward talukas (sub-regions) of Gujarat (S. Iyengar, 1988),

(b) The state organisations like National Bank could not gain credibility in supporting NGOs if they did not provide opportunity for exploration and experimentation to volunteers within their system,

(c) NGOs often did not recover even the operating costs of many services from people. In the process, such NGOs remained perpetually dependent upon aid agencies,

(d) Accountability of NGOs towards poor was rather low so much so that most NGOs did not aim at inducing poor people in their own management structures quite explicitly.

A nationalised bank invited the clerk to volunteer for two years village development work in an area of their choice without any loss of seniority I service on return. It triggered numerous innovative experiments by DVs.

The hands of DVs in technology generation, adaptation and diffusion system working on unpopular problems of larger social concern needed just as much attention? How to sustain professionals who disregard professional rewards and devote attention to such problems but can not put pressure for reform on their own organisations? Empowering them will require recognition of their voluntarism by a body of concerned scholars and activists. No national award has been given till date to any bank officer for initiating innovative scheme. So much so that about ten million rupees for new innovative schemes for rural development provided at national level remained unspent because no system existed for identifying and recognising DVs with in the mainstream system (Khanna, 1990).

How to resolve the paradox of developing rural poor's ability to manage their affairs by urban volunteers having inherent cultural limitations?

Poor do not cooperate with developmental organisations because we do not even acknowledge richness of their cultural and moral fiber. Findings of our research are seldom shared with those from whom we collect data (Gupta, 1987). Involving rural poor as co-researchers of social phenomenon, building upon cultural roots of voluntarism, respect for common property institutions can be invoked. Acknowledgement of local initiatives can provide spur for their transformation into innovations.

Documentation of people's knowledge and identifying scientific merit of some of the sustainable resource management alternatives can rekindle their experimental ethic. Institution building requires dispensability of external leadership, recognition of inverse relationship between status and skills and discrediting values generating helplessness. 'Lateral learning' among developmental volunteers as well as NGOs can be triggered to provide empirical basis of building theory-in/of action.

Search for culture specific metaphor and concept of voluntarism such as Zakat among Muslims, Gupt dan (anonymous charity) among Hindus, Kar Seva (voluntary labour for common good among Sikhs etc. are some examples of positive basis of different religions for building organic institutions. Different languages have words like "andi" (Haryanavi) and 'dhuni' (Hindi) implying a person obsessed with ideas generally for social good. Why has appreciation for this trait gone down? We submit that 'anonymous voluntarism' as a uniquely eastern concept has been subsumed under 'voluntary organisations'. The latter often believe that voluntarism can not exist in non-volunteer organisations. Should all innovations necessarily evolve into institutions? or should we aim at institutionalisation of culture which permits DBVs to explore the uncharted territories?

Finally, NGOs and developmental volunteers cannot succeed unless encapsulating and inhibitory values ingrained in the mind of poor during feudal historical past are questioned boldly.

Voluntarism in rural development in India has not been accompanied by pressure for policy change except around environmental issues. Often action at local level has not been linked with lobbying at macro level.

Recognising that state and markets perform better if kept under constant check, developmental volunteers within the organisations will have to perform a sort of 'insurgent' function so as to align, anonymously, with grassroot activities, NGOs and professionals.

International agencies can strengthen local social change by feeding not just the local ideas and innovations into global thinking but by providing global space for developmental volunteers to validate their hypothesis.

Right Livelihood Awards constitute one such source of international recognition. If rural poor of India could communicate with homeless in America, surely the cultures of deprivation will provide ground for collective action. Social innovators and DVs world over are struggling for the similar space in society where one does not have to go through a phase or unbridled accumulation followed by guilt, charity and benevolence for the have nots.

We doubt whether the term voluntary organisation by itself is much helpful. There are organisations carrying this label having no voluntary spirit amongst its members. At the same time there are volunteers in the public or market institutions who sometime perform insurgent function and sometimes extend the boundaries of responsibility of these organisations toward society.

The key concept is deviance and innovation. Space for providing scope for positive variance has to be created in each institution of society. It will be wrong to argue that all the individuals having voluntary spirit or an urge to go beyond their mandatory responsibility must move into NGOs or voluntary organisations. Absence of volunteers or socially committed professionals in the public or private organisations would make the task of NGOs nearly impossible.

Sustainability in nature and society requires players, whistle blowers, spectators, rules and the creative chaos. DVs are arguing that the losers in a game should not lose the right to play in the same field again. Asking them to play only in separate fields (in the form of volags) will eventually rob the game of the chaotic waves of sorrow and joy. Should we let it convert the spectators into warriors?

MINISTRY OF RURAL DEVELOPMENT

India has been a welfare state ever since her Independence and the primary objective of all governmental endeavors has been the welfare of its millions. Planning has been one of the pillars of the Indian policy since independence and the country's strength is derived from the achievement of planning. The policies and programmes have been designed with the aim of alleviation of rural poverty which has been one of the primary objectives of planned development in India. It was realized that a sustainable strategy of poverty alleviation has to be based on increasing the productive employment opportunities in the process of growth itself. Elimination of poverty, ignorance, diseases and inequality of opportunities and providing a better and higher quality of life were the basic premises upon which all the plans and blue-prints of development were built.

Rural development implies both the economic betterment of people as well as greater social transformation. In order to provide the rural people with better prospects for economic development, increased participation of people in the rural development programmes, decentralization of planning, better enforcement of land reforms and greater access to credit are envisaged.

Initially, main thrust for development was laid on agriculture industry, communication, education, health and allied sectors but later on it was realized that accelerated development can be provided only if governmental efforts are adequately supplemented by direct and indirect involvement of people at the grass root level. Accordingly, on 31st March 1952, an organization known as Community Projects Administration was set up under the Planning Commission to administer the programmes relating to community development. The community development programme inaugurated on October 2, 1952, was an important landmark in the history of the rural development. This programme underwent many changes and was handled by different Ministries.

In October 1974, the Department of Rural Development came into existence as a part of Ministry of Food and Agriculture. On 18th August 1979, the Department of Rural Development was elevated to the status of a new Ministry of Rural Reconstruction. That Ministry was renamed as Ministry of Rural Development on 23rd January 1982. In January 1985, the Ministry of Rural

Development was again converted into a Department under the Ministry of Agriculture and Rural Development which was later rechristened as Ministry of Agriculture in September 1985. On July 5, 1991 the Department was upgraded as Ministry of Rural Development. Another Department viz. Department of Wasteland Development was created under this Ministry on 2nd July 1992. In March 1995, the Ministry was renamed as the Ministry of Rural Areas and Employment with three departments namely Department of Rural Employment and Poverty Alleviation, Rural Development and Wasteland Development.

Again, in 1999 Ministry of Rural Areas and Employment was renamed as Ministry of Rural Development. This Ministry has been acting as a catalyst effecting the change in rural areas through the implementation of wide spectrum of programmes which are aimed at poverty alleviation, employment generation, infrastructure development and social security. Over the years, with the experience gained, in the implementation of the programmes and in response to the felt needs of the poor, several programmes have been modified and new programmes have been introduced. This Ministry's main objective is to alleviate rural poverty and ensure improved quality of life for the rural population especially those below the poverty line. These objectives are achieved through formulation, development and implementation of programmes relating to various spheres of rural life and activities, from income generation to environmental replenishment.

In order to ensure that the fruits of economic reform are shared by all sections of societies five elements of social and economic infrastructure, critical to the quality of life in rural areas, were identified. These are health education drinking water, housing and roads. To impart greater momentum to the efforts in these sectors the Government had launched the Pradhan Mantri Gramdoya Yojana (PMGY) and the ministry of rural development was entrusted with the responsibility of implementing drinking water, housing and rural roads component of PMGY

During the Ninth Plan period, several anti-poverty Programmes have been restructured to enhance the efficiency of the Programmes for providing increased benefits to the rural poor. Self Employment Programmes have been revamped by merging the Integrated Rural Development Programme (IRDP), the Development of Women and Children in Rural Areas (DWCRA), the Supply of Improved Tool-Kits to Rural Artisans (SITRA), the Training of Rural Youth for Self Employment (TRYSEM), the Ganga Kalyan Yojana (GKY) and the Million Wells Scheme (MWS) into a holistic self-employment scheme called Swarnjayanti Gram Swarozgar Yojana (SGSY).

Keeping in view the needs and aspirations of the local people, Panchayati Raj Institutions have been involved in the programme implementation and these institutions constitute the core of decentralized development of planning and its implementations. The Ministry is also vigorously pursuing with the State Governments for expeditious devolution of requisite administrative and financial powers to PRI's as envisaged under 73rd amendment act of the Constitution of India. On 25th December 2002, under Drinking Water Sector, a new initiative'Swajal Dhara' empowering the Panchayats to formulate, implement, operate and maintain Drinking Water Projects has been launched. In order to further involve PRIs in the development process, a new initiative'Hariyali' has been launched by Hon'ble Prime Minister on 27th January, 2003. Hariyali has been launched to strengthen and involve Panchayati Raj Institutions in the implementation of Watershed Development Programmes namely IWDP, DPAP and DDP.

The empowerment of rural women is crucial for the development of rural India. Bringing women into the mainstream of development is a major concern for the Government of India. Therefore, the programmes for poverty alleviation have a women's component to ensure flow of adequate funds to this section. The Constitutional (73rd) Amendment, Act 1992 provides for reservation of selective posts for women. The Constitution has placed enormous responsibility on the Panchayats to formulate and execute various programmes of economic development and social justice, and a number of Centrally Sponsored Schemes are being implemented through Panchayats. Thus, women Members and Chairpersons of Panchayats, who are basically new entrants in Panchayats, have to acquire the required skill and be given appropriate orientation to assume their rightful roles as leaders and decision makers. To impart training for elected representatives of PRIs is primarily the responsibility of the State Governments/Union Territory Administrations. Ministry of Rural Development also extends some financial assistance to the States/UTs with a view to improve the quality of training programmes and to catalyze capacity building initiatives for the PRI elected members and functionaries.

This Ministry is a nodal department for the two international organizations viz., the Centre on Integrated Rural Development of Asia and the Pacific (CIRDAP) and the Afro-Asian Rural Development Organization (AARDO).

The Ministry consists of the following three Departments:

1. Department of Rural Development
2. Department of Land Resources
3. Department of Drinking Water Supply

The Department of Rural Development implements schemes for generation of self employment and wage employment, provision of housing and minor irrigation assets to rural poor, social assistance to the destitute and Rural

Roads. Apart from this, the Department provides the support services and other quality inputs such as assistance for strengthening of DRDA Administration, Panchayati raj institutions, training & research, human resource development, development of voluntary action etc.

for the proper implementation of the programmes. The major programmes of the Department of Rural Development are Pradhan Mantri Gram Sadak Yojana, (PMGSY), Rural Housing (RH) Sampoorna Gramin Rozgar Yojana (SGRY) and Swaranjayanti Gram Swarozgar Yojana (SGSY).

Department of Land Resources implements schemes to increase the bio-mass production by developing wastelands in the country. Department also provides the support services and other quality inputs such as land reforms, betterment of revenue system and land records. It also undertakes development of desert areas and drought prone areas in the country.

The major programmes of the Department of Land Resources are Drought Prone Area Programme (DPAP) The Desert Development Programme (DDP) the Integrated Wasteland Development Programme (IWDP) and Land Reforms (LR).

These aim at increasing the soil and moisture conservation and productivity of the wasteland of the degraded lands thereby increase the income of the people. The provision of Drinking Water Supply and extension of Sanitation facilities to the rural poor are the main components of the activities of the Department of Drinking Water Supply. The major programmes of the Drinking Water Supply Department are The Swajaldhara, the Accelerated Rural Water Supply Programme (ARWSP) and the Total Sanitation Programme (TSP).

THE DISTRICT RURAL DEVELOPMENT AGENCY

Role and Functions

If effective programme design is critical to successful implementation of rural development programmes, so is an effective delivery agency. None of the anti- poverty programmes can have impact unless they are implemented with a clarity of purpose and a commitment to the task. It is here that the DRDAs play a critical role. The DRDAs are not be implementing agencies, but can be very effective in enhancing the quality of implementation through overseeing the implementation of different programmes and ensuring that necessary linkages are provided. To this extent the DRDA is a supporting and a facilitating organisation and needs to play a very effective role as a catalyst in development process.

The District Rural Development Agency is visualised as a specialised and a professional agency capable of managing the anti- poverty programmes of the Ministry of Rural Development on the one hand and to effectively relate these to the overall effort of poverty eradication in the District. In other words, while the DRDA will continue to watch over and ensure effective utilisation of the funds intended for anit-poverty programmes, it will need to develop a far greater understanding of the processes necessary for poverty alleviation/ eradication. It will also need to develop the capacity to build synergies among different agencies involved for the most effective results. It will therefore need to develop distinctive capabilities rather than perform tasks that are legitimately in the domain of the PRIs or the line departments. The role of the DRDA will therefore be distinct from all the other agencies including the Zilla Parishad.

DRDAs must themselves be more professional and should be able to interact effectively with various other agencies. They are expected to coordinate with the line departments, the Panchayati Raj Institutions, the banks and other financial institutions, the NGOs as well as the technical institutions, with a view to gathering the support and resources required for poverty reduction effort in the district. It shall be their endeavor and objective to secure inter-sectoral and inter- departmental coordination and cooperation for reducing poverty in the district. It is their ability to coordinate and bring about a convergence of approach among different agencies for poverty alleviation that would set them apart.

The DRDAs are expected to coordinate effectively with the Panchayati Raj The institutions. Under no circumstances will they perform the functions of PRIs.

The DRDAs are expected to oversee the implementation of different anti-poverty programmes of the Ministry of Rural Development in the district. This is not to be confused with actual implementation, which will be by the Panchayati Raj and other Institutions. The DRDAs will monitor closely the implementation through obtaining of periodic reports as well as frequent field visits. The purpose of the visit should be to facilitate the implementing agencies in improving implementation process, besides ensuring that the quality of implementation of programmes is high. This would include overseeing whether the intended beneficiaries are receiving the benefits under the different programmes.

The DRDAs shall keep the Zilla Parishad, the State and the Central Government duly informed of the progress of the implementation of the programmes through periodic reports in the prescribed formats. Special report, as and when called for, shall be provided.

It shall be the duty of the DRDAs to oversee and ensure that the benefits specifically earmarked for certain target groups (SC/ST, women and disabled) reach them. They shall take all necessary steps to achieve the prescribed norms.

The DRDAs shall take necessary step to improve the awareness regarding rural development and poverty alleviation particularly among the rural poor. This would involve issues of poverty, the opportunities available to the rural poor and generally infusing a sense of confidence in their ability to overcome poverty. It would also involve sensitizing the different functionaries in the district to the different aspects of poverty and poverty alleviation programmes.

The DRDAs will strive to promote transparency in the implementation of different anti- poverty programmes. Towards this end, they shall publish periodically, the details of the different programmes and their implementation.

Keeping in view, the substantial investment that are being made in poverty alleviation programmes, the DRDAs shall ensure financial discipline in respect of the funds received by them, whether from Central or State Governments. They shall also ensure that the accounts are properly maintained including in respect of the funds allocated to banks or implementing agencies in accordance with the guidelines of different programmes.

Thus the role of the DRDA is in terms of planning for effective implementation of anti-poverty programmes; coordinating with other agencies-Governmental, non-Governmental, technical and financial for successful programme implementation; enabling the community and the rural poor to participate in the decision making process, overseeing the implementation to ensure adherence to guidelines, quality, equity and efficiently; reporting to the prescribed authorities on the implementation; and promoting transparency in decision making and implementation.

In addition the DRDAs shall coordinate and oversee the conduct of the BPL Census and such other surveys that are required from time to time.

The DRDAs shall also carry out/ aid in carring out action research/ or evaluation studies that are initiated by the Central/ State Governments.

The DRDAs should deal only with the anti-poverty programmes of the Ministry of Rural Development. If DRDAs are to be entrusted with programmes of other ministries or those of the State Governments, it should be ensured that these have a definite anti-poverty focus. Entrusting of any programme to the DRDAs, other than anti- poverty programmes of the Ministry, be it of any other Ministry of Government of India or the respective State Government will have to be done with the approval of the Secretary, Rural Development of the respective State(s), who should examine such request in consultation with the Ministry of Rural Development, Government of India, In such cases, it must be ensured that adequate provision is made for requisite staffing needed for proper implementation of the programme.

Organizational Structure

In order to be effective, the DRDA must have an appropriate staffing structure as well as suitable personnel policy. Keeping in view the role and functions of the DRDA, the staffing structure of DRDA must include positions for Planning for poverty alleviation, Project formulation, Social organisation and Capacity building, Gender concerns, Engineering supervision and Quality control, Project monitoring, Accountancy and Audit functions as well as Evaluation and Impact studies. An indicative staffing structure is given in Annexure and is also detailed below. The State Government may modify the structure suitably, but without altering the basic design, to take care of the needs of individual districts keeping in view their size as well as specificity. However, this will also be subject to the overall ceiling of administrative costs admissible to the DRDAs in the State. By and large the staff appointed should be dedicated to DRDA- related works and should not be frequently transferred.

THE ROLE OF GOVERNMENT

Early Policy Developments

Many early postindependence leaders, such as Nehru, were influenced by socialist ideas and advocated government intervention to guide the economy, including state ownership of key industries. The objective was to achieve high and balanced economic development in the general interest while particular programs and measures helped the poor. India's leaders also believed that industrialization was the key to economic development. This belief was all the more convincing in India because of the country's large size, substantial natural resources, and desire to develop its own defense industries.

The Industrial Policy Resolution of 1948 gave government a monopoly in armaments, atomic energy, and railroads, and exclusive rights to develop minerals, the iron and steel industries, aircraft manufacturing, shipbuilding, and manufacturing of telephone and telegraph equipment. Private companies operating in those fields were guaranteed at least ten years more of ownership before the government could take them over. Some still operate as private companies.

The Industrial Policy Resolution of 1956 greatly extended the preserve of government. There were seventeen industries exclusively in the public sector. The government took the lead in another twelve industries, but private companies could also engage in production. This resolution covered industries producing capital and intermediate goods. As a result, the private sector was relegated primarily to production of consumer goods. The public sector also expanded into more services. In 1956 the life insurance business was nationalized, and in 1973 the general insurance business was also acquired by the public sector. Most large commercial banks were nationalized in 1969. Over the years, the central and state governments formed agencies, and companies engaged in finance, trading, mineral exploitation, manufacturing, utilities, and transportation. The public sector was extensive and influential throughout the economy, although the value of its assets was small relative to the private sector.

Controls over prices, production, and the use of foreign exchange, which were imposed by the British during World War II, were reinstated soon after independence. The Industries (Development and Regulation) Act of 1951 and the Essential Commodities Act of 1955 (with subsequent additions) provided the legal framework for the government to extend price controls that eventually included steel, cement, drugs, nonferrous metals, chemicals, fertilizer, coal, automobiles, tires and tubes, cotton textiles, food grains, bread, butter, vegetable oils, and other commodities. By the late 1950s, controls were pervasive, regulating investment in industry, prices of many commodities, imports and exports, and the flow of foreign exchange.

Export growth was long ignored. The government's extensive controls and pervasive licensing requirements created imbalances and structural problems in many parts of the economy. Controls were usually imposed to correct specific problems but often without adequate consideration of their effect on other parts of the economy. For example, the government set low prices for basic foods, transportation, and other commodities and services, a policy designed to protect the living standards of the poor. However, the policy proved counterproductive when the government also limited the output of needed goods and services. Price ceilings were implemented during shortages, but the ceiling frequently contributed to black markets in those commodities and to tax evasion by black-market participants. Import controls and tariff policy stimulated local manufacturers toward production of import-substitution goods, but under conditions devoid of sufficient competition or pressure to be efficient.

Private trading and industrial conglomerates (the so-called large houses) existed under the British and continued after independence. The government viewed the conglomerates with suspicion, believing that they often manipulated markets and prices for their own profit. After independence the government instituted licensing controls on new businesses, especially in manufacturing, and on expanding capacity in existing businesses. In the 1960s, when shortages of goods were extensive, considerable criticism was leveled at traders for manipulating markets and prices. The result was the 1970 Monopolies and Restrictive Practices Act, which was designed to provide the government with additional information on the structure and investments of all firms that had assets of more than Rs200 million, to strengthen the licensing system in order to decrease the concentration of private economic power, and to place restraints on certain business practices considered contrary to the public interest. The act emphasized the government's aversion to large companies in the private sector, but critics contended that the act resulted from political motives and not from a strong case against big firms. The act and subsequent enforcement restrained private investment.

The extensive controls, the large public sector, and the many government programs contributed to a substantial growth in the administrative structure of government. The government also sought to take on many of the unemployed. The result was a swollen, inefficient bureaucracy that took inordinate amounts of time to process applications and forms. Business leaders complained that they spent more time getting government approval than running their companies. Many observers also reported extensive corruption in the huge bureaucracy. One consequence was the development of a large underground economy in small-scale enterprises and the services sector.

India's current economic reforms began in 1985 when the government abolished some of its licensing regulations and other competition-inhibiting controls. Since 1991 more "new economic policies" or reforms have been introduced. Reforms include currency devaluations and making currency partially convertible, reduced quantitative restrictions on imports, reduced import duties on capital goods, decreases in subsidies, liberalized interest rates, abolition of licenses for most industries, the sale of shares in selected public enterprises, and tax reforms. Although many observers welcomed these changes and attributed the faster growth rate of the economy in the late 1980s to them, others feared that these changes would create more problems than they solved. The growing dependence of the economy on imports, greater vulnerability of its balance of payments, reliance on debt, and the consequent susceptibility to outside pressures on economic policy directions caused concern. The increase in consumerism and the display of conspicuous wealth by the elite exacerbated these fears.

The pace of liberalization increased after 1991. By the mid-1990s, the number of sectors reserved for public ownership was slashed, and private-sector investment was encouraged in areas such as energy, steel, oil refining and exploration, road building, air transportation, and telecommunications.

An area still closed to the private sector in the mid-1990s was defense industry. Foreign-exchange regulations were liberalized, foreign investment was encouraged, and import regulations were simplified. The average import-weighted tariff was reduced from 87 percent in FY 1991 to 33 percent in FY 1994. Despite these changes, the economy remained highly regulated by international standards. The import of many consumer goods was banned, and the production of 838 items, mostly consumer goods, was reserved for companies with total investment of less than Rs6 million. Although the government had sold off minority stakes in public-sector companies, it had not in 1995 given up control of any enterprises, nor had any of the loss-making public companies been closed down. Moreover, although import duties had been lowered substantially, they were still high compared to most other countries.

Political successes in the mid-1990s by nationalist-oriented political parties led to some backlash against foreign investment in some parts of India. In early 1995, official charges of serving adulterated products were made against a KFC outlet in Bangalore, and Pepsi-Cola products were smashed and advertisements defaced in New Delhi. The most serious backlash occurred in Maharashtra in August 1995 when the Bharatiya Janata Party (BJP—Indian People's Party)-led state government halted construction of a US$2.8 million 2,015-megawatt gas-fired electric-power plant being built near Bombay (Mumbai in the Marathi language) by another United States company, Enron Corporation.

AGRICULTURE

Agriculture has always been India's most important economic sector. In the mid-1990s, it provides approximately one-third of the gross domestic product and employs roughly two-thirds of the population. Since independence in 1947, the share of agriculture in the GDP has declined in comparison to the growth of the industrial and services sectors. However, agriculture still provides the bulk of wage goods required by the nonagricultural sector as well as numerous raw materials for industry. Moreover, the direct share of agricultural and allied sectors in total exports is around 18 percent. When the indirect share of agricultural products in total exports, such as cotton textiles and jute goods, is taken into account, the percentage is much higher.

Dependence on agricultural imports in the early 1960s convinced planners that India's growing population, as well as concerns about national independence, security, and political stability, required self-sufficiency in food production. This perception led to a program of agricultural improvement called the Green Revolution, to a public distribution system, and to price supports for farmers. In the 1980s, despite three years of meager rainfall and a drought in the middle of the decade, India managed to get along with very few food imports because of the growth in food-grain production and the development of a large buffer stock against potential agricultural shortfalls. By the early 1990s, India was self-sufficient in food-grain production. Agricultural production has kept pace with the food needs of the growing population as the result of increased yields in almost all crops, but especially in cereals. Food grains and pulses account for two-thirds of agricultural production in the mid-1990s. The growth in food-grain production is a result of concentrated efforts to increase all the Green Revolution inputs needed for higher yields: better seed, more fertilizer, improved irrigation, and education of farmers. Although increased irrigation has helped to lessen year-to-year fluctuations in farm production resulting from the vagaries of the monsoons, it has not eliminated those fluctuations.

Food-grain production increased from 50.8 million tons in fiscal year 1950 to 176.3 million tons in FY 1990. The compound growth rate from FY 1949 to FY 1987 was 2.7 percent per annum. Overall, wheat was the best performer, with production increasing more than eightfold in forty years. Wheat was followed by rice, which had a production increase of more than 350 percent. Coarse grains had a poorer rate of increase but still doubled in output during those years; production of pulses went up by less than 70 percent.

The increase in oilseed production, however, was not enough to fill consumer demands, and India went from being an exporter of oilseeds in the 1950s to a major importer in the 1970s and the early 1980s. The agricultural sector attempted to increase oilseed production in the 1980s and early 1990s. These efforts were successful: oilseed production doubled and the need for imports was reduced.

In the early 1990s, India was on the verge of self-sufficiency in oilseed production.After independence in 1947, the cropping pattern became more diversified, and cultivation of commercial crops received a new impetus in line with domestic demands and export requirements. Nontraditional crops, such as summer mung (a variety of lentil, part of the pulse family), soybeans, peanuts, and sunflowers, were gradually gaining importance.

The per capita availability of a number of food items increased significantly in the postindependence period despite a population increase from 361 million in 1951 to 846 million in 1991. Per capita availability of cereals went up from 334 grams per day in 1951 to 470 grams per day in 1990. Availability of edible oils increased significantly, from 3.2 kilograms per year per capita in FY 1960 to 5.4 kilograms in FY 1990. Similarly, the availability of sugar per capita increased from 4.7 to 12.5 kilograms per year during the same period. The one area in which availability decreased was pulses, which went from 60.7 grams per day to 39.4 grams per day. This shortfall presents a serious problem in a country where a large part of the population is vegetarian and pulses are the main source of protein.

There are large disparities among India's states and territories in agricultural performance, only some of which can be attributed to differences in climate or initial endowments of infrastructure such as irrigation. Realizing the importance of agricultural production for economic development, the central government has played an active role in all aspects of agricultural development. Planning is centralized, and plan priorities, policies, and resource allocations are decided at the central level.

Food and price policy also are decided by the central government. Thus, although agriculture is constitutionally the responsibility of the states rather than the central government, the latter plays a key role in formulating policy and providing financial resources for agriculture.

Land Use

In FY 1987, field crops were planted on about 45 percent of the total land mass of India. Of this cultivated land, almost 37 million hectares were double-cropped, making the gross sown area equivalent to almost 173 million hectares. About 15 million hectares were permanent pastureland or were planted in various tree crops and groves. Approximately 108 million hectares were either developed for nonagricultural uses, forested, or unsuited for agriculture because of topography. About 29.6 million hectares of the remaining land were classified as cultivable but fallow, and 15.6 million hectares were classified as cultivable wasteland.

These 45 million hectares constitute all the land left for expanding the sown area; for various reasons, however, much of it is unsuited for immediate cropping. Expansion in crop production, therefore, has to come almost entirely from increasing yields on lands already in some kind of agricultural use.

Topography, soils, rainfall, and the availability of water for irrigation have been major determinants of the crop and livestock patterns characteristic of the three major geographic regions of India—the Himalayas, the Indo-Gangetic Plain, and the Peninsula—and their agro-ecological subregions. Government policy as regards irrigation, the introduction of new crops, research and education, and incentives has had some impact on changing the traditional crop and livestock patterns in these subregions. The monsoons, however, play a critical role in determining whether the harvest will be bountiful, average, or poor in any given year. One of the objectives of government policy in the early 1990s was to find methods of reducing this dependence on the monsoons.

HIMALAYAN RANGES

The Himalayan region, with some 520,000 square kilometers of land, ranks well behind the other two regions in agricultural importance. Despite generally adequate rainfall, the rugged topography allows less than 10 percent of the land to be used for agriculture. The sandy, loamy soils on the hillsides and the alluvial clays in the region's premier agricultural subregion, the Vale of Kashmir—located in the northwestern part of the state of Jammu and Kashmir—provide fertile land for agricultural use.

The main crops are rice, corn, wheat, barley, millet, and potatoes. Most of India's temperate-zone fruits (apples, apricots, cherries, and peaches) and walnuts are grown in the vale. Sericulture and sheepherding also are being undertaken. In the eastern Himalayan subregion, the soils are moderately rich in organic matter and are acidic. Although much of the farming is done on terraced hillsides, there is a significant amount of shifting cultivation, which has resulted in deforestation and soil erosion. Rice, corn, millet, potatoes, and oilseeds were the main crops in the early 1990s. The region also is well known for the tea plantations of the mountainous Darjiling (Darjeeling) area in the northern tip of West Bengal.

Indo-Gangetic Plain

The vast Indo-Gangetic Plain, extending from Punjab to Assam, is the most intensively farmed zone of the country and one of the most intensively farmed in the world. Rainfall, most of which comes with the southwest monsoon, is generally adequate for summer-grown crops, but in some years vast areas are seared by drought. Fortunately, much of the land has access, or potential access, to irrigation waters from wells and rivers, ensuring crops even in years of drought and making possible a winter crop as well as a summer harvest. Wheat is the main crop in the west, rice in the east. Pulses, sorghum, oilseeds, and sugarcane are among other important crops. Mango orchards are common. Other fruits of the subregion include guavas, jackfruit, plums, lemons, oranges, and pomegranates.

In the Great Indian Desert, rainfall is scanty and erratic. About 20 percent of the total area is under cultivation, mostly in Haryana and Gujarat states, and comparatively little in Rajasthan. The Indira Gandhi Canal—begun in 1958 as the Rajasthan Canal—was designed to bring water from the north.

Progress was slow, and only the first stage was close to completion by the end of the Seventh Five-Year Plan (FY 1985-89). By then, the canal had substantially increased the area under cultivation in Rajasthan, and a new completion date of 1999 is anticipated. The cultivable area is expected to expand further with the development of the canal's second stage during the 1990s. The leading crops of the subregion are millet, sorghum, wheat, and peanuts. Vast expanses of sparse vegetation provide sustenance for sheep and goats. In the late 1980s, dairy farming became important in locations that had sufficient pastureland.

Peninsular India

The east and west coasts, the coastal plains, and the deltaic tracts that extend inland for some 100 to 200 kilometers in Peninsular India benefit from both the June-to-September southwest monsoon and the October-to-November northeast monsoon. Farther inland, as the topography and climate change, so does the pattern of agriculture. The proportion of land under cultivation ranges from about 50 percent along the coastal plain and in the western part of Andhra Pradesh to about 25 percent in eastern Madhya Pradesh. Except in areas of certain developed river valleys, double-cropping is rare. Rice is the predominant crop in high-rainfall areas and sorghum in low-rainfall areas. Other crops of significance along the east coast and in the Central Highlands in the early 1990s were pigeon peas, mustard, peanuts, millet, linseed, castor beans, cotton, and tobacco.

On the Deccan Plateau, deep, alluvial black soils that retain moisture for a long time are the basis for much of the region's output of farm products. However, the region also has many farming areas that are covered by thin, light-textured soils that suffer quickly from drought. Whether a crop is made or lost is, therefore, often dependent on the availability of supplementary water from ponds and streams. About 60 percent of the land in the state of Maharashtra was under cultivation in the early 1990s, less in Madhya Pradesh. About 75 percent of the cropland of the Deccan during this period was planted in food crops, such as millet, sorghum, rice, wheat, and peanuts; most of the remaining cropland was planted in fodder crops.

In the far south of the Peninsula, the area under cultivation varies from about 10 percent in the Western Ghats, to 25 percent in the western coastal tract, to 55 percent on the Karnataka Plateau. Here is the India—the land of spices—that Vasco da Gama and other European navigators came searching for in the fifteenth century. On the Karnataka Plateau, sorghum, millet, pulses, cotton, and oilseeds are the main crops on the 90 percent of the cultivated land that is dry-farmed; rice, sugarcane, and vegetables predominate on the

10 percent that was irrigated in the late 1980s. Coconuts, areca, coffee, pepper, rubber, cashew nuts, tapioca, and cardamom are widely grown on plantations in the Nilgiri Hills and on the western slopes of the Western Ghats.

Land Tenure

Matters concerning the ownership, acquisition, distribution, and taxation of land are, by provision of the constitution, under the jurisdiction of the states. Because of the diverse attitudes and approaches that would result from such freedom if there were no general guidelines, the central government has at times laid down directives dealing with the main problems affecting the ownership and use of land. But it remains for the state governments to implement the central government guidelines. Such implementation has varied widely among the states.

Landholding Categories

India is a land of small farms, of peasants cultivating their ancestral lands mainly by family labor and, despite the spread of tractors in the 1980s, by pairs of bullocks. About 50 percent of all operational holdings in 1980 were less than one hectare in size. About 19 percent fell in the one-to-two hectare range, 16 percent in the two-to-four hectare range, and 11 percent in the four-to-ten hectare range. Only 4 percent of the working farms encompassed ten or more hectares.

Although farms are typically small throughout the country, the average size holding by state ranges from about 0.5 hectare in Kerala and 0.75 hectare in Tamil Nadu to three hectares in Maharashtra and five hectares in Rajasthan. Factors influencing this range include soils, topography, rainfall, rural population density, and thoroughness of land redistribution programs.

Many factors—historical, political, economic, and demographic—have affected the development of the prevailing land-tenure status. The operators of most agricultural holdings possess vested rights in the land they till, whether as full owners or as protected tenants. By the early 1990s, there were tenancy laws in all the states and union territories except Nagaland, Meghalaya, and Mizoram. The laws provide for states to confer ownership on tenants, who can buy the land they farm in return for fair payment; states also oversee provision of security of tenure and the establishing of fair rents. The implementation of these laws has varied among the states. West Bengal, Karnataka, and Kerala, for example, have achieved more success than other states. The land tenure situation is complicated, and it has varied widely from state to state. There is, however, much less variation in the mid-1990s than in the postindependence period.

Independent India inherited a structure of landholding that was characterized by heavy concentration of cultivable areas in the hands of relatively large absentee landowners, the excessive fragmentation of small landholdings, an already growing class of landless agricultural workers, and the lack of any generalized system of documentary evidence of landownership or tenancy. Land was important as a status symbol; from one generation to the next, there was a tendency for an original family holding to be progressively subdivided, a situation that continued in the early 1990s. This phenomenon resulted in many landholdings that were too small to provide a livelihood for a family. Borrowing money against land was almost inevitable and frequently resulted in the loss of land to a local moneylender or large landowner, further widening the gap between large and small landholders. Moreover, inasmuch as landowners and moneylenders tended to belong to higher castes and petty owners and tenants to lower castes, land tenure had strong social as well as economic impact.

By the early 1970s, after extensive legislation, large absentee landowners had, for all practical purposes, been eliminated; their rights had been acquired by the state in exchange for compensation in cash and government bonds.

More than 20 million former zamindar-system tenants had acquired occupancy rights to the land they tilled. Whereas previously the landlord collected rent from his tenants and passed on a portion of it as land revenue to the government, starting in the early 1970s, the state collected the rent directly from cultivators who, in effect, had become renters from the state. Most former tenants acquired the right to purchase the land they tilled, and payments to the state were spread out over ten to twenty years. Large landowners were divested not only of their cultivated land but also of ownership of forests, lakes, and barren lands. They were also stripped of various other economic rights, such as collection of taxes on sales of immovable property within their jurisdiction and collection of money for grazing privileges on uncultivated lands and use of river water. These rights also were taken over by state governments in return for compensation. By 1980 more than 6 million hectares of waste, fallow, and other categories of unused land had been vested in state governments and, in turn, distributed to landless agricultural workers.

Land Reform

A major concern in rural India is the huge number of landless or near-landless families, many of whom are wholly dependent on a few weeks of work at the peak planting and harvesting seasons. The number of landless rural families has grown steadily since independence, both in absolute terms and as a proportion of the population. In 1981 there were 195.1 million rural workers: 55.4 million were agricultural laborers who depended primarily on casual farm work for a livelihood. In the early 1990s, the rural work force had grown to 242 million, of whom 73.7 million were classified as agricultural laborers. Approximately 33 percent of the employed rural workers were classified as casual wage laborers.

Because of the large number of landless farmers and the frequent neglect of land by absentee landlords in the early years of independence, the principle that there should be a ceiling on the size of landholdings, depending on the crop planted and the quality of the land, was embodied in the First Five-Year Plan (FY 1951-55). An agricultural census was conducted to provide guidance in setting such ceilings. During the Second Five-Year Plan (FY 1956-60), most states legislated fixed ceilings, but there was little uniformity among the states; ceilings ranged from six to 132 hectares. Certain specialized branches of agriculture, such as horticulture, cattle breeding, and dairy farming, were usually exempted from ceilings.

All the states instituted programs to force landowners to sell their over-the-ceiling holdings to the government at fixed prices; the states, in turn, were to redistribute the land to the landless. But adamant resistance, high costs, sloppy record keeping, and poor administration in general combined to weaken and delay this aspect of land reform. The delays in legislation allowed large landowners to circumvent the intent of the laws by spurious partitioning, sales, gifts to family members, and other methods of evading ceilings. Many exemptions were granted so that there was little surplus land.

To ensure more uniformity in income, to combat evasion of the intent of the laws, and to secure more land for distribution to the landless, the central government in the 1970s pushed for greatly reduced ceilings. For a family of five, the central government guidelines called for not more than 10.9 hectares of good, irrigated land suitable for double-cropping, not more than 10.9 hectares of land suited for one crop annually, and not more than 21.9 hectares for orchards. Exemptions were continued for land used as cocoa, coffee, tea, and rubber plantations; land held by official banks and other government units; and land held by agricultural schools and research organizations. At the option of the states, land held by religious, educational, and charitable trusts also could be exempted. To protect the states from legal challenges to their land reform laws, the constitution was amended in 1974 to include in its Ninth Schedule the state laws that had been enacted in conformance with national guidelines. Land reform laws enacted after 1974 also were included in the amendment.

By the beginning of the 1990s, all states and union territories, except Goa, Arunachal Pradesh, Nagaland, Manipur, Mizoram, and Tripura, had passed ceiling laws to conform to central government guidelines. In Maharashtra, for example, the revised ceiling law that became effective in 1975 set upper limits at perennially irrigated land, 7.2 hectares; seasonally irrigated land, 10.8 hectares; paddy land in an assured rainfall area, 14.6 hectares; and other dry land, 21.9 hectares.

By the early 1980s, about 150,000 hectares had been declared surplus under this act, about 100,000 of which had been distributed to 6,500 landless persons. A 1973 land reform amendment in Bihar set a range of ceilings on holdings for a family of five, from six to eighteen hectares depending on land quality, and offered an allowance for each additional family member, subject to a maximum of one-and-one-half times the holding. Within five years, the Bihar government had acquired 94,000 hectares of surplus land and had distributed 53,000 hectares to 138,000 landless families. Success nationwide was limited. Of the 2.9 million hectares of land declared surplus, nearly 1.9 million hectares had been distributed by the end of the seventh plan, leaving 1 million hectares still to be distributed as of early 1993.

By the early 1990s, nearly all the states had enacted legislation aimed at the consolidation of each tiller's landholdings into one contiguous plot. Implementation was patchy and sporadic, however. By the early 1980s, the work had been completed only in Punjab, Haryana, and western Uttar Pradesh and had begun in Orissa and Bihar. In most of the other states, nothing had been accomplished by the early 1990s. The Sixth Five-Year Plan (FY 1980-84) set a goal for the completion of the consolidation of holdings within ten years, which was not achieved.

In order to protect tenants from exorbitant rents (often up to 50 percent of their produce), the states passed legislation to regulate rents. The maximum rate was fixed at levels not exceeding 20 to 25 percent of the gross produce in all states except Andhra Pradesh, Haryana, and Punjab. The states also adopted various other measures for the protection of tenants, including moratoriums on evictions, minimum periods of tenure, and security of tenure subject to eviction on prescribed grounds only.

By the early 1980s, most of the cultivated area had been surveyed and records of rights prepared. In most states, revenue assessment—the tax on land—against farmland had been revised upward in keeping with a rise in farm prices. In several states, steps were taken to associate village assemblies, or *panchayat*, with the maintenance of land records, the collection of land revenue, and the management of lands belonging to government; the results of these efforts have frequently been unsatisfactory.

RURAL INDIA: CHALLENGES AND OPPORTUNITIES

Marketing today has completely transformed the nature and dynamics of business. One today needs to be adaptive to survive. Marketing of products has taken precedence over the process of production itself. This can be attributed to the fact that the new-age consumer equipped with the potent tool of information seeks more knowledge about the product, its features and its uses.

Customer today indeed is the "King". He can make or break the company. And when this information is presented in a creative and effective manner, it creates an everlasting impression on the consumer's mind and may even alter his perception of what he needs.

Marketers are forever seeking fresh challenges and scouting for more and more clientele to be drawn into their sphere of influence. The urban consumer has always been pampered with the most dazzling array of goods and services from every industry. But the urban market is fast shrinking due to saturation caused by the competition, and the growth rate over the past few years has consistently shown a declining trend. In the hunt for fresh pastures, the vast and hitherto vastly unexplored terrains of rural India consistently beckon the moolah-seeking marketer.

Why Rural India?

There are various reasons why every industry is taking a very serious look at rural markets: -

- About 285 million live in urban India whereas 742 million reside in rural areas, constituting 72% of India's population resides in its 6, 00,000 villages.
- The number of middle income and high income households in rural India is expected to grow from 80 million to 111 million by 2007 while urban India is expected to grow from 46 million to 59 million. Infact according to a recent survey, there are more *crorepatis* in rural Punjab than in a few big urban cities.
- Size of rural market is estimated to be 42 million households and rural market has been growing at five times the pace of the urban market.
- More rural development initiatives by the government.
- Increasing agricultural productivity leading to growth of rural disposable income.
- Lowering of difference between taste of urban and rural customers.
- Good Monsoons in the last couple of years.
- Growing rural infrastructure - thanks to Government initiatives.

- Setting up of channels like e-choupals by companies like ITC.

Many companies like Colgate-Palmolive, HLL, Godrej, etc., have already made forays into rural households but still capturing the markets is a distant dream. Most marketers still lack in-depth knowledge to analyze the complex rural market.

In the Indian context, rural marketing is a complex subject. For a business organization, rural marketing is beset with a number of problems. The prices of rural marketing pose many problems due to the vastness of the country and a high potentiality for providing an effective marketing system.

Besides, a few other problems stem from the under-developed markets, and illiterate and gullible people constitute the major segment of the markets. More purchasing power is not enough.

It is not enough to have some consumption pioneers. The activation of buying on a wide scale is an essential pre-condition for the exploitation of the rural market. It is now unanimously accepted that the rural salesmanship in India has been insufficient and inadequate and out of proportion to the agriculture revolution. This calls for strong bias in favour of raising the rural demand as against the urban demand.

The traditional marketing activities of promotion, distribution, sales and servicing, undertaken so far in the urban and semi-urban contexts, are to be extended to cover a much wider area in a rural environment by introducing appropriate innovation, selection and adoption.

There are about 5,76,000 villages in India, 79 percent of them with a population of less then 1,000 each. The major hurdles in tapping the rural markets can be summarized as: -

1. High distribution costs
2. High initial market development expenditure
3. Inability of the small retailer to carry stock without adequate credit facility
4. Generating effective demand for manufactured foods
5. Wholesale and dealer network problems
6. Mass communication and promotion problems
7. Banking and credit problems
8. Management and sales managing problems
9. Market research problems
10. Inadequate infrastructure facilities (lack of physical distribution, roads warehouses and media availability)
11. Highly dispersed and thinly populated markets

12. Low per capita and poor standards of living, social, economic and cultural backwardness of the rural masses
13. Low level of exposure to different product categories and product brands
14. Cultural gap between urban-based marketers and rural consumers

The development of the rural market will involve additional cost both in terms of promotion and distribution. In rural marketing, often it is not promotion of a brand that is crucial, but creating an awareness concerning a particular product field, for instance, fertilizers and pesticides. Urban and semi-urban based salesmen are not able to tap the full potential in the villages. Here, it may be suggested that the marketers may select and employ the educated unemployed from villages.

METHODS FOLLOWED TRADITIONALLY

Traditional methods of rural marketing make an interesting study and they ought to be analyzed carefully to draw relevant conclusions. Conventionally, marketers have used the following tools to make rural inroads:

- Use of few select rural distributors and retailers to stock their goods but no direct interaction with prospective consumer.
- Use of print media or radio but no alternate form of advertising for promoting their brands.
- More focus on price of product but less attention devoted to quality or durability.
- Same product features for urban and rural setting with no customization for rural areas despite differences in the market environment.
- Low frequency of marketing campaigns.
- Little uses of village congregations like *haats* and *melas* to sell the products.
- More focus on men as decision makers and buyers.

Strategies for Rural Marketing

The past practices of treating rural markets as appendages of the urban market is not correct, since rural markets have their own independent existence, and if cultivated well could turn into a generator of profit for the marketers. But the rural markets can be exploited by realizing them, rather than treating them as convenient extensions of the urban market. The focus should be on infecting marketing culture into the villages. The educated unemployed youth in the villages could be trained to carry out this mission.

1. Decentralizing rural markets by detaching them from the urban bases. A give-and-take two-way approach should replace the present one-way exploitation.
2. The salesman in rural markets should be selected from the educated unemployed villagers, trained well and appointed as salesmen. The town-to-villages shuttling salesmen are to be replaced by stationary salesman in villages.
3. Companies should also adequately concentrate on educating the villagers to save them from spurious goods and services.
4. Rural markets are laggards in picking up new products. This will help the companies to phase their marketing efforts. This will also help to sell inventories of products out dated in urban markets.

Future Outlook

Considering the magnitude of the task at hand with the companies, it makes sense for non-competitive companies like HLL and LG to make a joint effort to penetrate the market. They can use each other's distribution channels to leverage their brands. Also considering the poor awareness levels of the people, competitors like HLL and P&G should join hands to avoid the product proliferation, which results in confusing the consumer. Rural consumers have a very high level of ethos so all the care should be taken not to hurt them in any form of advertising. Moreover, every effort should be made by the companies to promote the "my brand" feel in the minds of consumers. This can be achieved by connecting the local industries of that place and, if possible, use it in packaging or graphics of the product. So the need of the hour is the integration of all the above stated approaches and constantly reinventing the products and bringing new innovations. Companies will need to think beyond the short-term gains and resort to experimentation so as to develop a dedicated model for these areas. This will ensure rich rewards and long-term association with rural consumers.

CO-BRANDING - BEYOND JUST BRANDS

Onboard Singapore-Airlines, you will notice a special smell. The smell called Stefan Floridan Waters is a branded Singapore-Airlines smell patented by the airline and sprayed on all the hot towels given to passengers. Microsoft has for years used a start-up sound.

There are obvious reasons for the obsession to occupy a secure place in the consumers' mind. Brand Coke is currently valued at $ 69.5 billion against Coke's worldwide revenues of $ 29.9 billion. Companies invest enormous amounts of time and money in building brands. But today's media-bombarded consumers hardly have enough attention to spare. Even if they do have, they are armed with the means to skip all advertisements. And then, there is the problem of diminishing customer loyalty due to a proliferation of brands.

Co-branding evolved as one possible solution to the above problems. It is a smart way for companies to come together and offer consumers a new product with the blend of their respective expertise. Enhanced product features, quality satisfaction and the pride of being associated with multiple brands at the same time, have done the job of attracting more consumers. Other advantages like increased brand visibility, cost-cutting and easy diversification have prompted companies the world over to get into this game.

Different brands can come together to lend different flavors to a unique final product. Flavors are distinctive brand attributes. Think of the unique taste that you get when you eat a plain Vanilla Ice Cream on which Pepsi has been poured.

Co-Branding: Marriages are Good!

Physicists have a term amplification reserved for the phenomena behind the working of transistors and lasers. Management specialists have a word synergy to convey more or less the same effect. Biologists prefer symbiosis. But then, how often is the equation 1+1=3? For that matter, can the equation actually hold?

The IBM-Intel, Diet Coke-Nutrasweet, Sony-Kodak products are some classic examples that have proved the above point. All of these products have created enormous value for the stakeholders involved. Co-branding is a meeting of expertise and reputation of individual partners; the combination of whom creates a whole new product/service and occupies a unique position in the minds of consumers.

Branding was the initial step that companies used, to differentiate themselves and attract the 'knowledge rich' consumers to make repeat purchases and increase their bottom lines. But even differentiation has reached its limit and the ever-demanding consumer's attention is scarce today. Co-branding gives companies a wider playing field to differentiate their products and give their consumers a whole new experience, while keeping the core brand values same.

Promotional co-branding, ingredient co-branding, value chain co-branding and innovation based co-branding have contributed to companies' kitties to varying degrees. But now, marketers need to go a step further in order to grab the customers' attention. It is here that 'brand flavouring', a concept that has been introduced, will give marketers that extra mindshare of the consumers.

'Brand flavouring' requires a careful amalgamation of two or more brands, wherein the overall product has a unique appeal, while the participating brands' values are still the same. It is like the unique taste that you get when you have an ice cream with a gulab jamun or an ice cream with Pepsi. Note that the individual tastes are still retained here. Add a few nuts to the above combination and the experience can get even better!

Co-Branding in India

Co-branding is now increasingly featuring in the marketing strategies of many Indian companies. Companies have realised that smart co-branding can help them boost their brand image, improve sales through sharing of distribution networks, enhance customer loyalty, increase customer base, enter new segments and garner a host of such advantages with little extra expenses.

Co-Branding - Marry & Make Merry

Aggressive marketing companies have sparked off a new trend of purchasing, by issuing co-branded credit cards with banks and financial institutions. Co-branded credit cards from LG and SBI, ICICI and HPCL, Air-Sahara and Standard-Chartered Bank, HSBC and Star India Bazaar, show that these have spread across to all possible business sectors in India. The popularity of these cards in India is on the increase as the 'price conscious' consumers get to enjoy extra exclusive benefits from multiple recognised brands at no extra cost. Co-branded credit cards spur spending as they reward loyal customers with rewards, encashable bonus points, discounts, cash-back offers and a host of such schemes. Ease and convenience during shopping, payment of bills and transacting at petrol or retail outlets are other reasons. For the card issuers, it is an easy way to directly expand their consumer base and increase customer loyalty. For the other partner, there is an access to a database with complete details of a large number of customers which can make their segmentation and targeting of customers much more systematic, thus increasing their profitability.

According to an ICICI estimate, about 30 per cent of the credit cards issued by it are of the co-branded variety. This is because most of the new issues are co-branded since they are actually stimulating sales.

Marry the Right Person

Co-branding can be an interesting way for companies to venture into new areas. Boots-Piramal and Saregama-Records are jointly producing a series of music albums of old Hindi songs. Boots-Piramal fits in perfectly in this alliance as Strepsils from Boots Piramal has always been marketed on the 'voice' platform. Boots-Piramal has taken Strepsils from the 'smooth voice' platform to a next level of 'sweet voice' platform by its arrangement with Saregama. Both of them will also co-promote each other's products at their respective outlets, and thus increase their brand visibility. Co-branding has helped MTV India enter new territories apart from television programming. The MTV brand has lent its youthful 'funky' character to Citibank credit cards, Airtel SIM cards and Fastrack watches. MTV has had reasonable success in its co-branding ventures with Airtel SIM cards and Citibank credit cards. Here, the participants cooperate because they have, or want to consciously achieve, an alignment of their brand values in the customer's mind. As is evident, Airtel and Fastrack have re-asserted their youthful character with their MTV partnership.

Marriages are Made in Heaven - Keep Them Good & They'll Do You Good

P&G, India, undertook a co-branding exercise with the National Association for Blind in the form of Project Drishti, where one rupee per pack of Whisper purchased by the customer was diverted towards the cause of a blind female child. Using the funds so collected, P&G arranged for the restoration of eye-sight of two hundred and fifty blind girls through corneal transplant operations. This deal gave P&G much more than it invested. By associating itself with a noble cause, it came to be associated in the public mind with a worthy cause.

Cause-related marketing is a form of co-branding. The 1983 partnership between American Express and the project to restore the Statue of Liberty demonstrated the benefits associated with an effective cause-related marketing strategy. The result of the campaign was a 28% increase in AMEX card use, with a 17% rise in new card applications, while raising $1.7 million for the Statue of Liberty repairs.

Corporations need to pursue this strategy effectively as this will boost their brand image tremendously in today's world of Corporate Social Responsibility. Co-branding in this way can change consumers' perceptions about the companies' products or services, and increase employee satisfaction. Combining their operational expertise with the human resources and knowledge of an NGO, companies can achieve both individual and social goals.

Brand Flavoring - The Next Step

Brand flavoring is about presenting the picture in a different way. Building and nurturing new brands takes a lot of time and effort. Also, at a time when product life cycles have shortened tremendously and lost meaning, creating new brands at the same pace is definitely not an option. It is here that brand flavouring can be an option.

Brand flavoring is a kind of extension of co-branding, wherein the attempt is to bring diverse partners together. Different brands can come together to lend different flavors to a unique final product. Flavors are distinctive brand attributes. The strength of such an exercise will depend on how well the partner brand values can contribute to the overall product/service. The concept is limited in the sense that only strong brands can really contribute in such an environment without diluting their individual brand equity.

Using this concept, I have presented some ideas, some highly ambitious ones, but whoever said that anything was impossible. (Some of them are pure co-branding solutions, though.)

- Of the people... for the... AIR has an approximate coverage of 92% of the Indian sub-continent while Doordarshan reaches approximately 90% of the population, which is far more than the combined reach of all the satellite channels put together. Yet, no co-branding arrangements exist between public and private enterprises.
- Co-branding will help media houses share content and technology, increase their household reach and revenues and do much more to their image. More than just increasing revenues, such arrangements will increase the quality of content and do social good
- Helping Hand: The Indian SME sector's contribution to GDP was about 40% in 2004 and over 11.4 million SSI units provided employment to about 27.1 million people. Also, SME's contributed to about 40% of the country's industrial production and 34% of the country's exports.

Now, the twist in the story. Rane Brakes Linings Ltd. demonstrated its world class quality by winning the Deming award in 2003.

Yet, even today only 3% of its total revenues come from overseas exports. Co-branding with auto majors like TATA, Hyundai or Toyota, will help it improve its brand image and market share in the global market.

This is just a single example. Co-branding arrangements are possible between Arvind Brands, Versace and an established cotton supplier in Coimbatore. What about a local food processor tying up with Amul or Haldirams? How about extending the alliance of one with Heinz or Kraft?

Co-branding in such a way will help local companies develop a global image. In this age of SCM and CRM, co-branding must be used to increase operational efficiency and profitability. All this will finally lead to a huge growth in GDP, as the SME's are major players in our economy.

- Brand India: Thailand and Malaysia being smaller than MP attract more foreign tourists than the whole of India!! We now have an "Incredible India" brand, but that is not enough. Private-public co-branding is very essential here. Which foreign tourist will not be lured by an Incredible India package co-branded with a hospitality partner in Taj-Hotels, an airliner in Jet-Airways, a travel partner in Cox & Kings, and a wireless mobile-cum-internet provider in Airtel?
- Taking this to a next level, how about co-branding arrangements with SAARC or ASEAN countries? How about a business trip arranged in Singapore, followed by a business agreement signed in Hyderabad, and then a relaxing evening in Phuket? Why should we lose $ 900 million annually, due to lack of co-branding arrangements between India and Pakistan?
- While in Rome... At a time when Indian companies are making a mark on the world stage, smaller companies will find co-branding a very convenient tool to capture new markets. Titan, being a world class company, is still facing a lot of problems in expanding into the European market. Would co-branding with Swatch or Tissot be an alternate solution? It always helps to be-friend a Roman while in Rome. On the same lines, ethnic Tanishq jewellery co-branded with Swarowski would definitely lure any European lady.
- Of Superpowers and more... How about a convergence enabled product from Lenovo of China, TCS & Airtel of India and NTT DoCoMo of Japan? For all this to happen, the talent pool needs to be developed. We need to foster partnerships in education where students learn the best from an IIT in India, Harvard in USA and an NTU in Singapore. Think of the brand appeal of these students! This is very important, as India's large talent pool needs to be well tapped and encouraged, for India to become a superpower.

2

The Rural Community and Economic Development

The approach taken by Community Economic Development practitioners relies on a distinctive blend of principles, strategies, and tactics.

PRINCIPLES OF COMMUNITY ECONOMIC DEVELOPMENT

In an effort to build a theory of CED and to establish best practices, a group of CED practitioners formed the Working Group on CED in British Columbia, Canada in 1991. As a result of a one-year consultation they came up with the following guiding principles:

CED is an evolving, on-going process based on equity; participation; community-building; cooperation and collaboration; self-reliance and community control; integration; interdependence; respect for ecological limit; capacity-building; diversity; and the use of appropriate indicators.

Strategies and Tactics of Community Economic Development

A number of strategies and tactics have been identified to help ensure that the guiding principles of CED are implemented. CED practitioners' efforts concentrate on:

- Addressing the primary needs of all community members through; sustainable supply of nourishing food and safe drinking water, security and safety of persons and property, adequate and affordable housing, opportunity for health and recreation, increased quality of community life through mutual support, and opportunities for meaningful work.
- Developing a strong and sustainable local market economy through; local economic planning process, keeping money circulating in the community and stopping leakage, encouraging enterprise in the community, building local capacity to sustain CED, and developing means to access local capital to finance CED initiatives.

- Strengthening civil society and increasing civic participation through; citizenship, cultural development, responsible government, community infrastructure, participatory democracy, and public involvement.
- Creating and supporting opportunities for community learning through; education and skills development, sharing knowledge of locality, opportunities to learn through doing, participatory action research, engaging elders in sharing wisdom and experience, and creating computer networks,.
- Strengthening natural ecosystems by; increasing awareness of bio-regionalism and ecology, planning for a sustainable environment, reducing exploitation of resources, engaging in environmental restoration, developing and supporting appropriate technology, creating land trusts for nature and natural resources, and establishing a community forest.
- Strengthening the social / cultural economy through recognition, validation and structural support by; valuing, encouraging and supporting volunteerism, setting up a skill exchange program, creating supports for housework, creating support for home health care, and supporting cultural diversity and development.

RURAL ECONOMY

The success story of Indian economy is spreading its wings towards India's districts and rural areas which accounts for over 65% of the nation's total populace. Currently, the rural sector of India is not only observing a massive increase in its per capita income but also in its expenditure and production. To enhance the rural economy of India, the Indian government has increased the monetary incentives allotted for the National Rural Employment Guarantee Act (NREGA) to USD 8.04 bn in its 2009 Union Budget. Moreover, the government has also allotted USD 34.74 bn in its Bharat Nirman Programme for enhancing rural road and rail network. Surprisingly, the rural economy was unaffected by the recent global financial meltdown as per the research conducted by Rural Marketing Association of India (RMAI). The research also revealed that the rural economy in India is soon to witness an increase in its earnings which will be largely triggered by the incessant expansion in agriculture for the last four successive years.

RURAL ECONOMIC ENVIRONMENT

The Indian economy has been passing through the trajectory of high growth for several years. The overall Gross Domestic Product (GDP) is estimated to grow at 8.4 per cent, with the agriculture and allied sector projected to bounce back with 3.9 per cent growth during 2005-06. The Government has also been consolidating its fiscal position, creating a conducive climate for investment activities. Investment demand in the economy is picking up, reflecting the growing dynamism in the industrial sector.

The exports have also assumed vibrancy. Thus, the Indian economy as a whole is poised for still higher growth in the coming years. The Rural Economy in India is wholly agriculture based and it is of tremendous importance because it has vital supply and demand links with the other Indian industries. Agriculture is the main stay of the Indian economy, as it constitutes the backbone of rural India which inhabitants more than 70% of total Indian population.

The fertility of the soil has augmented the success of agriculture in India. Further, Rural Economy in India has been playing an important role towards the overall economic growth and social growth of India. India has been predominantly an agriculture-based country and it was the only source of livelihood in ancient time. During prehistoric time when there was no currency system the India economy system followed barter system for trading i.e. the excess of agricultural produce were exchanged against other items. The agriculture produce and system in India are varied and thus offers a wide agricultural product portfolio.

Today, the rural economy in India and its subsequent productivity growth is predicated to a large extent upon the development of its 700-million strong rural population. The agricultural economy of India is drafted according to the needs of rural India since majority of the population lives in about 600,000 small villages. In India, agriculture accounts for almost 19% of Indian gross domestic products (GDP). The rural section of Indian population is primarily engaged with agriculture, directly or indirectly. The Ministry of Agriculture, the Ministry of Rural Infrastructure, and the Planning Commission of India are the main governing bodies that formulate and implements the policy related to rural economy in India and its subsequent development for the overall growth of the Indian economy.

The main agricultural products that controls the fate of the Rural Economy in India are as follows:-

- *Food Grains:* Rice, Wheat, Pulses, Cereals, Corn, Maize, Rice Bran Extractions, Sorghum, Soy meal, Suji, Parmal, Lentils, Jowar, Bajra, Chick pea
- *Fruits and Nuts:* Cashew Kernels, Cashew Nut, Cashews, Almonds, Roasted Dry Fruits, Peanuts, Groundnut, Walnut Kernels, Walnuts, Indian Peanuts, HPS Groundnuts

- *Fruits:* Bananas, Beans, Cherry, Cucumbers, Dried Fruits, Dried Truffles, Carrots, Lemon, Mandarins, Mango steens, Meslin, Shallots, Apples, Asparagus, Grapes, Oranges, Gherkins, Turnips, Oranges, Papaya, Pineapple
- *Vegetables:* Potatoes, Bitter gourd, Stripe Gourd, Pumpkin, cauliflower, Cabbage, Tomato, Onion, Green Pepper, Drum Sticks, Lady's finger, Banana, Papaya, Spinach, Cucumber, Mushroom, Mushroom Spawn, Radiata
- *Seeds, Buds, Plantation and Related Products:* Basil Seed, Cumin seeds, Dill Seed, Buds, Celery Seed, Hybrid Seeds, Sesame Seeds, Sesbania Seed, Sunflower Seeds, Mustard Seeds, Oil Seeds, Plant Products, Plantation, Plants, Psyllium Seed, Fennel Seed, Fenugreek Seed, Herb Seeds, Tamarind Seed, Vegetable Seeds
- *Spices:* Black Pepper, Chilli Powder, Chillies, Cinnamon, Cloves, Coriander Powder, Cumin, Curry Powders, Dry Ginger, Dry Red Chilly, Cardamom, Anise, Salt, Onion Powder, Pepper, Fenugreek, Clove, Ginger, Turmeric, Turmeric Powder
- *Tea and Coffee:* Black Tea, Coffee, Coffee Beans, Darjeeling Teas, Assam Teas, Instant Coffee, Leaf Coffee, Leaf Tea, Packaged Tea, Green Tea, CTC Teas
- *Tobacco and Tobacco Products:* Beedi, Betel nut Leaves, Betel nut, Bidi Leaves, Chewing Tobacco, Cigarettes, Arecanut, Jarda, Scented Tobacco, Smoking Tobbacco, Snuff, Opium, Pan, Chatni, Pan Masala, Gutkazarda, Zafrani Zarda
- Cotton, Rubber, Jute etc.

No. The uneven growth will never help India to grow faster or help achieving the projected results. I support the PURA model of President A P J Abdul Kalam which calls for rural development and uplift of BPL with a social cause.

Finance is still the biggest problem for development of rural economy despite it being innovative. People in the rural areas have no money and often tend to go for high-cost borrowings. No venture capitalists or funders have ever come forward to help those community with social obligations in mind.

Who lends the money to those poor people without collateral securities despite having the entrepreneurial capabilities, innovativeness, creativeness and hard working abilities? It is a pity to say that in few districts where I visited, villagers happened to be bonded labourers to big money launderers due to high cost borrowings which has never been a easy task to repay. The development of the Indian Rural Economy can be credited to the liberal policy of the Ministry of Rural Development, Government of India and the Planning Commission of India.

India is still an agriculture-based country and the agrarian sector contributes substantially to the GDP of India. Around seventy percent of the Indian population still depends on agriculture and allied areas for livelihood.

- Glimpses of Rural Economy of India
- A target of Rs 225,000 crores for farm credit has been set for the financial year 2007-2008.
- 50 lakh new farmers have been brought under the banking system.
- Agricultural Insurance to facilitate agricultural loans to the farmer.
- Allocation for the Rural Infrastructure Development Fund to be increased substantially.
- In the financial year 2006-07, 35 projects were successfully completed.
- Additional irrigation of 900,000 hectares has been targeted in the financial year 2007 - 2008.
- For better farm yields modern farming techniques has been set up.
- Direct disbursement of subsidy to rural Indian farmers has been arranged through a pilot programmeme.

Role of Regional Rural Banks

The importance of the rural banking in the economic development of a country cannot be overlooked. As Gandhiji said "Real India lies in villages," and village economy is the backbone of Indian economy. Without the upliftment of the rural economy as well as the rural people of our country, the objectives of economic planning cannot be achieved. In fact, the real growth of Indian economy lied in the emancipation of rural masses from acute poverty, unemployment, and socio-economic backwardness.

Keeping this end in view, various important plans and programmememes of rural development have been conceived and implemented by the government of India since the commencement of first five-year plan from 1951-56. But an appraisal of the achievement of these programmememes clearly reveals that much programmememes failed to achieve the desired objectives due to the backward economic condition and lack of adequate finance to the poor people in the rural areas. Hence, bank and other financial institutions are of vital importance for development of rural economy of a country.

The present study is a modest attempt to make an appraisal of the credit needs of the rural people and the way Regional Rural Bank, i.e., Arunachal Pradesh Rural Bank, has been extending its service to meet the same in the state of Arunachal Pradesh. It deals with the performance evaluation of Arunachal Pradesh. Rural Bank (APRB) for the economic development of the state. Further, an attempt has also been made to study the growth and performance of Scheduled Commercial Banks with special emphasis on Regional Rural Banks (RRBs) in Indian and North-East Region.

A New Rural Economy

It is very possible that villagers are investing a part of their small surpluses in non-farm enterprises. Are there some subterranean changes in rural India that get ignored in the wider debate on farmer distress? The provisional results of the economic census that was conducted in 2005 were released.

They suggest that there is a gradual increase in non-farm activity in our villages. Take two facts. First, the average annual increase in the number of rural enterprises since 1998 was 5.53%. New urban enterprises cropped up at a slower pace of 3.71%. Second, since 1998, average annual employment growth in rural India was 3.33%, compared with 1.68% in the cities.

What does this mean? It is very possible that villagers are investing a part of their small surpluses in non-farm enterprises. If this is indeed the case, then what we are seeing are the first signs of an overdue diversification of the rural economy. The new units are offering employment opportunities to young villagers. Farming still dominates, but watch this trend as well.

Through the past four years, India's Gross Domestic Product (GDP) has grown at a phenomenal rate of 8.6%. The momentum has continued through the current fiscal as well. The challenge today, however, lies as much in sustaining this momentum as ensuring an inclusive pattern of growth. I expect two distinct thrust areas in the coming Budget: a focus on social and physical infrastructure, and initiatives to integrate the rural economy with the national economy.

Since shortage of skilled manpower constitutes a serious long-term challenge to the economy today, Indian industry would expect a concerted effort from the finance minister on the education and skill development platform. In line with the 11th Plan Approach paper, which prescribes allocation towards education be raised from the current 2.9% to 6% of GDP, I expect enhanced allocation towards creation of a robust education and skill development infrastructure in the country.

We look forward to initiatives towards development of an institutional framework to devise and implement national, sectoral strategies to develop and improve the skills of Indian workforce. Such a move would help us graduate from a labour to a knowledge arbitrage economy.

Given that over-stretched physical infrastructure is going to be a major concern area, Indian industry would expect significant initiatives in the area of resource mobilisation for infrastructure projects. The country has by and large developed a consensus around accelerated inclusion of the rural economy in the wider growth process. We expect added thrust on two critical components of the rural economy: strengthening of irrigation infrastructure and a larger role for the private sector to improve agricultural productivity and rural income. The government needs to attract more private companies towards agricultural R&D through appropriate tax and financial incentives.

On the tax front, Indian industry would expect some reduction in overall tax burden and removal of tax-related anomalies to stimulate manufacturing. In the service sector, telecom would expect dilution in the overall incidence of various levies, which cumulatively stand at a staggering 30% of operator revenues. Such an initiative would help operators penetrate further into the rural hinterland and enable telecom to play a key role in the integration of the rural to the national economy. On the whole, Corporate India expects a Budget that not only drives economic growth but ushers in the key instruments for equitable growth.

Higher Rural Allocation

Apart from providing loan waivers to more than four crore small and marginal farmers in the country, the Union Budget has essentially focused on revamping the rural agricultural Economy through various measures such as augmenting rural credit, increasing the resource base of banks that are into priority sector lending and increasing allocation to social sector schemes under the National Rural Employment Guarantee Act (NREGA) and Bharat Nirman.

The agricultural credit of schedule banks, which is estimated to around Rs 2.4 lakh crore during the current fiscal, would go up by Rs 2.8 lakh crore during 2008-09, marking a significant 16.6% hike.

The finance minister has also proposed allocation of Rs 10,200 crore for increasing the resource base of financial institutions such as National Bank for Agriculture and Rural Development (Nabard), Small Industries Development Bank of India (Sidbi) and National Housing Bank (NHB) to meet the requirement of priority sector lending. "Financial inclusion can be carried forward by expanding the reach of these three banks," Union finance minister P Chidambaram said. A fund of Rs 5,000 crore in Nabard would enhance its refinance operations to short-term cooperative credit institutions, he said. There are proposals for two funds of Rs 2,000 crore each for risk capital financing and enhancing refinance capability to micro, small and medium enterprises. Besides, a fund of Rs 1,200 crore in NHB would also be created to enhance its refinance operations in the rural housing sector.

The Gross Capital Formation in agriculture as a proportion of GDP in the agriculture sector has improved from 10.2% during 2003-04 to 12.5% during 2006-07. The government targets to raise it to 16% during the eleventh plan for achieving a growth rate of 4% in the agriculture sector. It implies that more investment in the agriculture sector is in the offing during the eleventh plan period. According to the Budget, more than Rs 1,185 crore had been released to the four states following the recommendation of Vaidyanathan committee on reviving the short-term cooperative credit structure being implemented across 17 states.

There has been an agreement between the states and the Centre for implementation of the committee report on reviving the long-term cooperative credit structure. The Centre's share will be Rs 2,642 crore or 86% of the total burden. "All this thrust on the rural Economy would definitely ensure that farmers return to the farming," says Richard Mahapatra, coordinator, Centre for Science & Environment, a New Delhi based... NGO. He added that the agriculture sector is not only the biggest employer in the country, but also plays a crucial role in sustaining the rural Economy.

Besides, a higher allocation of Rs 16,000 crore for NREGA—the Union government's flagship social programmememe—is expected to go a long way in reviving the agriculture sector as most of the recommended works relate to water conservation. "Some of the key initiatives such as expansion of NREGA scheme to 596 districts in the country will have a positively impact and create livelihood options for rural youth in horticulture, biofuels, etc," says Rana Kapoor, CEO, Yes Bank. The Budget also announced reduction on duty on vitamin premixes and mineral mixtures from 30% to 20% with the intention of reducing the production cost of cattle and poultry feeds. The duty on phosphoric acid has been reduced from 7.5% to 5%, while duty on bactofuges is reduced from 7.5% to nil. This will boost the dairy industry.

Rural Economic Development

Nobel Prize winning economist Robert Lucas had remarked that once you start thinking about economic growth, it is hard to think of anything else. What are the causes of economic growth and how can the process be enabled is a question that has obsessively occupied some of the best minds in the world of economics and commerce.

The question takes on an unparalled urgency and importance when applied to the rural Indian economy because it presents an enormous challenge, and, consequently, presents an equally great opportunity for making a difference in the lives of hundreds of millions of people. India's economic development is predicated upon India's rural development because around

700 million Indians live in rural India. An astonishing one out of every ten living humans lives in rural India.

Rapid progress in GDP growth and globalization in the last decade has primarily impacted the urban economy. While software exports, business process outsourcing, etc, have helped urban economic growth, it has done relatively little for the rural economy. Without rural economic development, India has little chance of achieving growth rates required to become a developed nation. Furthermore, economic development is both a cause and a consequence of urbanization.

Clearly, in the Indian context, urbanization through further rural to urban migration is both unsustainable and socially disruptive. Therefore urbanization of the rural population will have to be achieved in the rural areas. Rural India is caught in what is called a development trap. Because of lack of economic opportunities, incomes are low. Therefore they are unable to pay for goods and services that would enable them to increase their incomes. This leads to low demand for goods and services. Consequently, firms don't find it profitable to do business in rural India. This leads to the inadequate provision of infrastructure, which in turn leads to lack of economic opportunities, and so on.

It is important to recognize that human capital is the scarce resource globally. Fortunately India is lavishly endowed with immense human capital.

However, physical capital is in relatively short supply in India. The challenge therefore is to use the limited capital most efficiently to break out of the poverty trap by integrating the rural economy with the urban Indian economy and indeed the global economy.

Various models for rural economic growth have been proposed and implemented. Vinod Khosla and I have proposed a model which harnesses the power of the information and communications technology (ICT) revolution to accelerate rural economic growth. The model called Rural Infrastructure & Services Commons (RISC) has the potential for achieving the multi-faceted goals of sustainable development. It uses limited resources efficiently by focusing them in specific locations that are accessible to a sufficiently large rural population, such as that of 100 villages. RISC provides the benefits of urbanization by making available to rural populations the full set of services and amenities that are normally available in urban areas.

It brings the benefits of ICT and the increased access to global markets that globalization promises. The model recognizes that rural populations face a number of inter-related gaps, not just the celebrated digital divide. Bridging them simultaneously with a holistic solution is more likely to succeed than any partial intervention can. The model facilitates the coordination of the investment decisions of the private sector, the public sector, NGOs, and multilateral lending institutions.

To achieve its goal, the model strikes a number of balances — between the local and the global, between planned infrastructure investment and market-driven service provision, between specialization and standardization. It does not require government subsidies for its continued operation, although the government does have a role in providing some critical functions such as risk alleviation, loan assistance, and enacting enabling legislation.

A typical RISC installation would provide services for about 100,000 rural people. These services, mostly but not all provided competitively by a large number of for-profit firms, will range from education, health, market making, financial intermediation, entertainment to government services, social services, etc.

Since all services themselves require the infrastructural services such as power, telecommunications, water, physical plant, etc., large specialized firms will provide the infrastructure. RISC obtains urbanization economies, which arise from the agglomeration of populations and infrastructure facilities. By installing RISCs to serve the rural populations of an entire state, economies of scale and scope are also obtained. Scale economies would be significant at each level of the model.

At the infrastructure level, there are transaction costs associated with the necessary coordination between the firms providing the core infrastructural services. At the services level, the cost of the services will be inversely proportional to the quantity demanded and supplied. A RISC provides a complete set of services and functions. Each service provider itself is a customer of other services co-located on the RISC. The banker uses the internet and postal services, and the internet service provider uses the banking and postal services, and so on.

They make each other mutually viable and even possible. All these economies essentially lower the cost of service provision and, in a competitive market, makes them more affordable. At a certain level of abstraction, the proximate causes of poverty can be seen as two gaps: the ideas gap and the objects gap. The objects gap is the lack of physical resources – too little land, too little capital stock, etc – that contribute to persistent poverty. The ideas gap is the lack of knowledge about how to make the best use of the resources available.

Fortunately, the cost of knowledge goods has dropped precipitously due to the revolution in information and communications technologies. Bridging the ideas gap is a much easier task than ever before. RISC uses ICT intensively towards that end. The transition from the concept to the actual implementation of the RISC model requires co-ordination of investment decision of the government and the large firms that provide the infrastructural elements. It is a non-trivial but surmountable challenge provided the political will and the vision exists among policy makers, private sector leaders, leading investors, and opinion makers.

Urbanization and Rural Economy

One should avoid following the straight jacketed approach followed by our smartest of all political leaders to comment on such issues. A 'Neta' (Perhaps we need to change the meaning of this word in Hindi dictionaries now) vouching for rural votes will say 'No' and someone vouching for urban votes will say 'Yes'.

An interesting and intriguing incident has been the downfall of BJP government, a government riding high on 'India Shining' campaign. The downfall has been associated with observations that BJP was concentrated on Urban India only. The truth of the analysis is evident form the fact that the campaign was in worded in English was pointed towards BJP's urban India approach. Ahem!

The Congress therefore followed their slogan as 'Aam Aadmi ki sarkar', worded in Hindi and worded for rural people. They made their huge plans {I have deliberately used the word 'Plans'} to associate with the rural people. Such developments may lead to conclude "Yes - urbanization has hindered the growth of rural people". In fact, I have also been to an extent convinced with this philosophy though I have thought that why concentrate on one side, why cannot both sides develop simultaneously.

- A study of consumption in last 26 years reveals that an increase in Urban Household consumption of Rs 100 leads to an increase of Rs 39 in the rural household income. The driver of such increase is non farm sector of rural economy such as manufacturing, construction and trade, hotels and restaurants.
- During the last decade, the rural economy is estimated to have grown by 7.3% as compared to 5.4% of urban economy.
- The rural economy is nearly as big as the urban. In 2000, the rural economy accounted for 49% of the India's GDP.
- Share of agriculture in Rural economy has dropped from 72% in 1970s to 64% in 1980s to 51% in 2000
- The rural-urban migration has fallen from 6.5% in 1981 to 2.8% in 2001.

The indications pointed by each point noted above leads to refute a number of myths such as rural people are suffering due to urbanisation; there is an India – C waiting to shine and many more - an encouraging realization!!!!

The major reason for this change, as noted by eminent researcher and economists is the growing and strengthening integration of the rural economy with the urban economy. Granted that logistics and supply chain for the route is no less than a shame for the progress we hope to achieve, yet the hope does exist. Perhaps an integration of the rural supply chains with the urban ones, steered by the boom in organised retail would provide the impetus to pass on the benefits lower down the pyramid. Just one problem though – Who will help Mayawati understand on the recent Reliance Fresh issues? Anyone ready to bell the cat?

Winds of Growth

There may finally be a silver lining to the gloomy economic picture prevailing in the country. Last week, there were economic forecasts from three important quarters that saw a better future for the country's economy in the current year. The Centre for Monitoring Indian Economy (CMIE) forecast a 6.3 per cent growth in real GDP in the current financial year against 5.2 per cent growth in GDP in the last fiscal.

Its forecast was quickly followed by estimates from the Reserve Bank of India (RBI) which saw the economy growing at over 6.5 per cent in FY2001-2002 and then the finance minister stated that he saw the economy turning around by the end of the calender year. In a nutshell, the picture for Indian economy appears to be brightening up as we go forward into the year.

However, the most remarkable feature of this improving economic outlook is that the growth is projected to come almost entirely from the agriculture sector. CMIE has forecast that growth in the agriculture production during the current year would be a whopping 9 per cent over the previous year. This would be almost the highest in the last seven years and comes after two consecutive years of shrinkage in agriculture sector.

Impact on Rural Economy

After having grown at over seven per cent annually during 1994 and 1997, India's GDP growth slowed down substantially because of the sharp fall in agricultural production and income. After an impressive 8.2 per cent rise in production in 1998-99, the pace of growth in the farm sector turned negative with output shrinking by 1.5 per cent in FY2000 and 4.5 per cent in FY2001.

This meant there was less disposable income generated in the rural areas and hence the demand pull from vast majority of population was absent.

The last two years were marked by a drought in states like Gujarat, Maharashtra and Rajasthan which contribute significantly to domestic production and also account for substantial amount of consumer spending.

With a pick up in income in these areas, the demand for goods and services is expected to pick up this year and that should contribute to the over all growth in economy.

Impact on Industry

Increased income in farm sector impacts the entire economy, however it has an almost immediate impact on some sectors like consumer durables and non-durables, tractors and motorcycles. There is also an increased demand for cement and steel from the resurgent housing sector. Heavy commercial vehicles too attract buying as transportation of goods picks up across the country and demand for services too is higher.

Therefore, the relationship between agricultural income and economic growth is almost direct. We expect FMCG companies, tractor manufacturers and motorcycle makers to feel an immediate pick up in demand once the kharif crop is harvested. This sentiment has already been reflected in the stock markets with the FMCG leader, Hindustan Lever attracting huge amount of buying by fund managers in expectation of a large top line growth this year.

We might also see a turnaround in the subdued sentiment in industries such as televisions, washing machines and refrigerators which have seen a fall in sales due to sluggish rural demand over the past two years. Higher demand for goods and services in rural areas necessitates increased movement of goods and hence a pick up in demand for commercial vehicles. This means increased demand in almost all sectors such as steel and engineering. Therefore, if the agriculture sector does grow at the projected 9 per cent this year, economy would receive a big boost by the end of the year.

Growth in Industry

The growth in industry suffers from poor demand in consumption as well as investments. Due to a decline in farm income over the past two years, demand from rural areas was poor and as result of this investment sentiment remains sluggish too. The situation can only reverse if consumption rises giving a boost to investment sentiment. However, as the impact of increased income in farm sector would be felt only from the end of the year, the growth in industrial production in expected to remain this year.

CMIE has projected that the growth in industrial production this year would be nearly 4.5 per cent, down from 5 per cent in the last financial year and the lowest level since 1998-99. The CMIE forecast notes that the index of industrial production (IIP) registered a growth of 5 per cent last year and 6.6 per cent in 1999-2000. While in April 2001 industrial production grew only

2.7 per cent and in May it was even lower at 1.9 per cent, the overall growth may be better as the impact of improved demand is felt later in the year.

Morgan Stanley Dean Witter too has projected that the industrial activity will pick up in the second half of this calendar year and the IIP should be between 4 and 5 per cent for the year.

Gains from Rains

According to the CMIE study, the 2001 south west monsoon has been fairly widespread and intensive across most parts of the country. Between June 1 and June 27, 2001, as many as 29 out of the 35 meteorological sub-divisions in the country received excess to normal rains, covering almost 90 per cent of the gross cropped area. Out of this 75 per cent of the area received excess rains.

In comparison, in June 2000 only 72 per cent of the cropped area in 26 meteorological sub-divisions had received excess rains, the CMIE report states. Thus rains have not only been good they have also been evenly distributed leading to prospects of widespread growth. Rains have been particularly good for Eastern and Western UP, Haryana and Punjab, raising hopes of a huge demand rise in these states early net year. The only areas that have received deficient rainfall so far include Rayalseema in Andhra Pradesh, interiors of Karnataka and some parts of Tamil Nadu.

Early and good rains have enthused farmers to go for kharif sowing operations ahead of the normal schedule. The CMIE report says that it expects kharif crop to be much better this year than the harvest of 2000.

Stocks to Watch

As we have stated earlier, the direct beneficiary of a surge in rural demand would be companies in the FMCG sector, like Hindustan Lever. HLL has seen a sluggish topline growth over the pat two years largely because of a slowdown in demand from rural areas. The company can be expected to capitalise heavily on a revival in farm income. The markets have already anticipated this and last week several prominent funds were buying into the counter. Infact, Hindustan Lever was one of the biggest gainers in the stock market index last week. Others that could gain too include Nirma, Colgate Palmolive, Britannia Industries and Tata Tea.

The other gainers could be tractor majors who have faced a sharp slump in demand over the past two years. The decline in farm income had forced a big slump in tractor sales. Tractors sales are almost first to be impacted by a change in farm income. We expect that tractor majors such as Punjab

Tractors and Mahindra & Mahindra would be biggest beneficiaries of a look up in the prospects. Both the stocks, particularly, Mahindra & Mahindra, has taken a big hit in valuations due to the fall in sales. Both the companies are fundamentally sound and have very good management. We could see a major re-rating of the industry in the coming months. Investors could watch out for these stocks.

Also expect motorcycle sales to pick up and companies Hero Honda Motors could be the biggest gainers. While other two-wheeler makers should also see a pick up in demand, Hero Honda can outperform the market by a long margin. Other manufacturers of consumer durables like colour televisions, refrigerators and washing machines could also see a surge in demand driven by consumption in rural economy. Manufacturers such as Electrolux Kelvinator, Whirlpool and Voltas could see demand rising for their products. Electrolux and Whirlpool could be particularly fancied due to their large capacities and strong brand presnce.

A surge in rural economy would also mean more transportation needs, increased housing and larger demand for services. CMIE projects the growth in services at 7 per cent in the current year. We also expect commercial vehicle makers like Telco and Ashok Leyland to gain. Cement and steel could also see increased demand from a pick up in housing activity.

Overall, we may say that the worst may soon be over for the economy and things could begin to look much better from the second half of the year.

Rural India is nothing but agriculture; and since agriculture has been growing at a mere 1.9 per cent since 1995-96, so is the rural economy. Therefore, per capita rural income has fallen in the recent past. "We have a dual economic structure in which a 1.9 per cent growth economy supports more than 60 per cent of the population, and a 7.3 per cent growth economy supports the remaining 40 per cent".

The rural-urban chasm is not only deep but also "has grown substantially over the past few years". For cell phones, other durables, consumer goods and services, penetrating the rural market will be limited by poverty and lack of growth, and that "companies might find themselves reaching saturation point much sooner than expected."

This gives a misleading picture of the economy, and would perhaps have been understandable if it had been a populist politician that had equated the rural economy with agriculture to say that two-third of India's population live on only 22 per cent of the country's income, while a third live in cities and lord over 78 per cent of the GDP.

Since April last year, we have constructed what is probably the most comprehensive picture of India's rural economy — a digitised database of economic, demographic, consumption data that has been extracted and harmonised across various sources, goes back over two decades, down to 530 districts. All the evidence that we have collected, processed and analysed tells us that rural is much more than agriculture. This is true not just for the present, but has been so for quite some time. In 1993-94, the Central Statistical Organisation conducted its last rural-urban classification of India's net domestic product (NDP, which is GDP minus depreciation).

It showed that rural India accounted for 54 per cent of the country's NDP, while agriculture contributed to 31 per cent of India's GDP. So, even if one were to account for depreciation, the rural economy of 1993-94 was quite a bit more than just agriculture, forestry and fishing. The share of rural economy is more or less the same seven years later. Our analysis shows that in 2000-01, out of India's NDP of Rs 1,062,400 crore (Rs 10,624 billion) at constant 1993-94 prices, the share of rural India was 52 per cent.

This is how rural NDP played out in 2000-01: agriculture accounted for 46 per cent; industry took up another 21 per cent; and services 33 per cent. Between 1993-94 and 2000-01, rural NDP at constant prices grew at an average of 6.2 per cent per year. Since agriculture grew at around 2 per cent during this period, clearly the rural economy has been much more than just agriculture. There is a lot of other evidence as well that demonstrates the growth of non-agricultural activities throughout rural India. According to the Annual Survey of Industries, even in 1993-94, rural India accounted for 29 per cent of the country's organised manufacturing units; 30 per cent of its employees; 32 per cent of its output; and 30 per cent of its net value added.

In 2000-01, over a rising base, rural India could speak of 36 per cent of organised manufacturing establishments, 38 per cent of its employees, 43 per cent of output; and 41 per cent of net value added. Consider also the Census of India. In 1991, 30.6 per cent of rural households had permanent (or "pucca") houses. By 2001, this had risen to 41 per cent, over a base that was growing at 1.7 per cent a year. Between 1991 and 2001, households using LPG for cooking rose from 1 per cent to almost 6 per cent, with Himachal Pradesh leading at 22 per cent. In 2001, almost a fifth of rural households owned TV sets, and the rural regions of 10 states were significantly above the national average.

Between 1991 and 2001, the percentage of rural households with electricity connections rose from 30 per cent to 43.5 per cent and 11 states ranked well over the national rural average. In 2001, over 30 per cent of rural households kept their funds in banks or postal savings accounts. All this doesn't square up with the age-old image of rural India — that of highly indebted, underfed peasants ploughing handkerchief-sized fields under the blazing heat of the mid-day sun. That may be still be partly true of agriculture; but not of the rural economy. We won't inundate you with more evidence. Instead, let's make three points. One, there's more to rural than just agriculture; it has a thriving and growing manufacturing and services sector. Two, between 1993-94 and 2000- 2001 per capita rural income grew at the same pace as urban income. The reason why total urban income was greater has much to do with higher population growth. Three, there is considerable pent-up demand for goods as well as services such as education, telecom, medical treatment and financial products.

Those in the top quartile of rural India have been discontinuously higher spenders than the average, spending at least Rs 180,000 crore (Rs 1,800 billion) — NSS, 55th round — in 1999-2000, with expenditure patterns that are pretty similar to urban India. Consider this: services can't constitute 52 per cent of GDP on the rump of only the urban populace.

Reconstruction of the Rural Economy

Skill and enterprise will take them part of the journey. But they cannot bring back the lost prosperity unless the entire economy of the province, now severely deficit in several directions, is placed on stable foundations. In any programmememe of reconstruction, means of irrigation take the first place. Four-fifths of the new allottees had holdings even in West Punjab in the lowest grade of ten standard acres and less.

At least a third of the evacuee land in East Punjab and Pepsu has gone to the small rural traders from the predominantly Muslim districts of West Punjab. Many of them may take to the land, but theirs will continue to be an impoverished economy, for the call on their traditional skills cannot but be meagre. In virtue of the land held by them they have obtained land now, and this will give each of them a home and a location, but not as a general rule an adequate living.

Individuals from this group of settlers need assistance and encouragement to leave their families at home and to venture out into other parts of the country, where their intelligence and resourcefulness will enable them to build a new composite way of life intertwined with their interests in their own province. In this way they can re-establish themselves. Tenants from West Punjab as well as those who cultivated for large Muslim holders in East Punjab present another important problem of reconstruction.

Altogether, reduced holdings and poorer lands, a fall in the number of actual workers on the soil, insufficient irrigation facilities, slender capital resources, and the rise of tenant interests, combine to produce a picture of increasing economic pressure, which the land may not long sustain. If the stresses are not to overwhelm the social economy itseif, the end of the resettlement operations should be followed by a period of bold administrative action, reform in the management of land, and investment of capital on a scale sufficient to produce results. The migration has brought to East Punjab a peasantry uncommonly rich in experience and progressive in outlook, whose record of achievement over sixty years of colonization is second to none in the world. In this and in the spirit of enterprise, which raised the western districts of the Punjab to the highest level of rural prosperity ever, reached in the long history of India lies perhaps the best hope of the future.

The East Punjab Government propose to replace the present system of temporary allotments of evacuee lands by a new system of allotments which will take account of the holdings of evacuees in West Punjab. The new allotments will not confer rights of ownership or permanent occupancy, but the possession of allottees will be maintained. Claims of allottees will be dealt with in accordance with decisions reached eventually regarding the treatment of evacuee property.

In the new scheme of allotments, land will be allotted only to those who, In West Punjab, were owners, occupancy tenants under the Punjab Tenancy Act, and tenants under the Colonization of Government Lands Act and to certain other classes of grantees and holders of land in West Punjab to be of equivalent areas, while in the case of larger holders there will be graded specified by Government. It is proposed to give to small holders allotments cuts.

The definition of the " Small Holders " and the details of the graded cuts will be determined when detailed information regarding the available areas in East Punjab and the East Punjab States, the areas held by the popu-lation to be settled in East Punjab and the East Punjab States, and other relevant information becomes available. It is intended to complete the new system of allotments in East Punjab and the East Punjab States, not later than the 31st May, 1948. Governments are, however, anxious to introduce the new scheme as early as may be feasible and steps to this end will be taken at once.

Arrangements for collecting complete information regarding the land available for allotment in East Punjab and the East Punjab States and the land abandoned by individual evacuees will be taken in hand without delay and it is hoped also to make arranagements on a reciprocal basis to secure Information from records of rights in West Punjab. To ensure accurate information an Ordinance will shortly be promulgated prescribing punishment lor false information regarding claims to land and action by way of forfeiture and otherwise in respect of allotments taken on false information.

Claims to land will be invited on a form to be prescribed by Government. Until the new system of allotments can be introduced, the present system of allotments will continue and allotments made to the present holders will be maintained subject to a complete scrutiny of existing allotments, cancel-lation of unauthorised and excessive allotments, dispossession from illicit occupation and such other adjustments as may be necessary Including ad-justments in the unit of allotment decided upon by Government. Government desires that kharif crops including sugar-cane and cotton should be sown to the maximum extent.

Accordingly they wish to assure the present allottees that in the event of their being otherwise dispossessed before the kharif crops have been harvested. Government will make arrange-ments for compensation after proper verification, through sharing of the crop or in some other appropriate manner. 7th February, 1948.

Land Claims Procedure

The purpose of this chapter is to outline in some detail the steps taken and the instructions laid down from time to time in working on land claim' applications submitted by about half a million individuals. This description is given for two reasons. For several years it will be necessary for revenue officials to refer to papers connected with the allotment of land to displaced persons and to check whether the entries were according to the instructions and also, in some particulars, what the precise object of the various entries was. Secondly, the circumstances under which work on land claims was done during 1948 and 1949 imposed certain difficulties.

Not all the difficul-ties could be anticipated and, as the work proceeded, new problems arose. The experience gained in dealing with large numbers of claims to property and the errors which came to light in the course of this work should be recorded, so that when similar work is under-taken in other fields, mistakes and omissions can be avoided. It is possible in retrospect to distinguish between features, which arose from the nature of the task handled, and those, which were due to local or particular factors.

Inviting Land Claims

In the press note, which was issued on the 7th February, 1948 with the concurrence of the East Punjab States and the Government of India, the East Punjab Government announced its intention to invite claims to land abandoned by displaced persons. The East Punjab Refugee (Registration of Land Claims) Ordinance, 1948 was thereupon promulgated and corresponding legislation was introduced by the States of Patiala, Kapurthala. Nabha, Jind and Faridkot. The ordi-nance provided definitions of expressions such as "claim", "land", "land-holder", "refugee" and "registering officer". Claims were to be invited under the Ordinance from refugee landholders in respect of land situated in West Punjab, N.W.F.P., Sind, Baluchistan and States adjacent to these provinces, which had acceded to the Dominion of Pakistan, in other words, from displaced landholders from Western Pakistan.

The expression, "refugee" was defined in the Ordinance as:

"A landholder in the territories now comprised in the Provinces of West Punjab, N.W.F.P., Sind or Baluchistan or in any State adjacent to any of the aforesaid Provinces and acceding to the Dominion of Pakistan, who has since the 1st day of March, 1947, abandoned or been made to abandon his land in the said territories on account of civil disturbances, or the fear of such disturbances, or the partition of the country."

This definition was amended by the legislature before passing the East Punjab Refugees (Registration of Land Claims) Act. 1948 to mean a— "landholder in the territories now comprised in the Province of West Punjab, or who or whose ancestor migrated as a colonist from the Punjab since 1901 to the Provinces of North-West Frontier Province, Sind or Baluchistan or to any State adja-cent to any of the aforesaid Provinces and acceding to the Dominion of Pakistan, and who has since the 1st day of March, 1947, abandoned or been made to abandon his land in the said territories on account of civil disturbances, or the fear of such disturbances, or the partition of the country."

On the 7th April, 1948 the relative responsibilities of the Central and the Provincial Governments for the rehabilitation of dis-placed persons were considered in the Joint rehabilitation Board, in which both Governments were represented. The record of pro-ceedings sets out the discussion and the decision in respect of settle-ment on evacuee land: "After the East Punjab Government have made up their mind in re-gard to the refugees they are going to rehabilitate in the East Punjab and East Punjab States, the rehabilitation of the rest in the States of Bikaner, Bharatpur and Alwar will be the responsibility of the Centre. After some discussion, it was decided that the East Punjab Government will undertake the responsibility of rehabilitation in East Punjab and East Punjab States, of Sind. N.W.F.P., Bahawalpur and Baluchistan refugees who are of Punjabi extraction- Other refugees from Bahawalpur will be rehabilitated in Bikaner and else-where. Refugees from Sind, N.W.F.P. and Baluchistan who are not of Punjabi extraction will be rehabilitated outside East Punjab and East Punjab States, e.g., Bharatpur, Alwar etc."

Penalties

No authentic records being available, it was felt that if severe penalties were provided, dishonesty in the submission of claims could be greatly reduced.

The Ordinance provided rigorous imprisonment upto five years or fine upto Rs. 5000 or both as penalties for persons submitting false claims or a public servant abetting such an offence-In addition to these penalties, Deputy Commissioners were authorised to disqualify from allotment by an order passed in writing any person who has submitted a false claim and to cancel, either in whole or in part, any allotment which may have been made on the basis of a false claim. An appeal from such an order lies within thirty days to the Financial Commissioner whose decision is final and conclusive and cannot be called in question in any court.

The Claim Form

By rules made under the Ordinance, the East Punjab Govern-ment prescribed a form on which claims for land were to be submitted. The East Punjab Government's form was used also in the States. The form was sold at the price of one anna. It was printed in Urdu, Punjabi and English. It provided for an affidavit by the applicant, signature or thumb-impression by one wit-ness and attestation by a competent authority. The rules under the Ordinance prescribed competent authorities for the purpose of attest-ing affidavits on the land claim application.

Lacuna In the Form

For every district in which land was abandoned by a dis-placed person, one claim form was to be filled. Details of holdings in different villages of the same district were to be entered on one form. Subsequently, in actual work, since the particulars of land held by a person had to be verified and completed village by village, supplemen-tary land claim forms were prepared in the Land Claims Office.

Copying of extracts from claims which were frequently incomplete or badly written on such an extensive scale led to mistakes in supplemen-tary forms. Again, when it became necessary to consolidate into a single account, land held by an individual in different villages in more than one district, it was a serious handicap to have begun with claim forms on a district basis.

Experience showed that the right course would have been to have obtained a separate form for each single vil-lage in which an individual held land along with a covering form giving particulars of all villages together, whether situated in one district or in several. The form did not specially provide space and directions for entering an applicant's address. This omission was soon noticed and the necessary instructions were issued to registering officers. Direc-tions to applicants for intimating changes in their addresses should also have been given. A column in the form asked for particulars of village, tehsil and district of origin from applicant who was colonists from different districts. The word "colonist" was not defined.

Gradually, in the course of land claims work, the word has come 'o mean not one necessarily who goes out from a particular district and "colonizes" land, but one who belongs to a particular district, whether of East Punjab or of West Punjab, who held land in another district. Since the claims of colonies from West Punjab were to be consolidated with their West Punjab districts and the claims of colonists from East Pun-jab and the States were to be sorted out for settlement in their parent districts, the column came to assume increasing importance. Incomplete information in this column led later to many difficulties in con-solidating individual accounts.

Arrangements for Registration

The Registration of Land Claims Ordinance provided for "registering officers," a term which included revenue officers within their jurisdiction, a Land Claims Officer at the central office and spe-cially appointed officials known as Assistant Land Claims Officers. Claims applications were to be presented under the rules to registering officers either in person or through recognised agents or they were to be sent by registered post. In the case of minors, applications had to be submitted by their next friends or guardians. An application sub-mitted by post unaccompanied by an affidavit not attested by a compe-tent authority was not to be entertained.

Arrangements for registration were made at the offices of tehsils and sub-tehsils throughout East Punjab and East Punjab States. Each tehsil carried its own classification number. Annexure I give a complete statement of classification numbers which were adopted. A receipt was given for every claim received. Although the period 10th March, to the 31st March, 1948 was prescribed as the period for the registeration of land claims, the date was extended upto the 10th April, and even later small numbers of claims continued to arrive by post.

The work of registration was conducted satisfactorily. Nevertheless, it was felt that if more time had been given and claims had been written by trained hands there would have been a greater measure of accuracy.

Work in the Land Claims Office

The press note of the 7th February, 1948 had announced the intention of Government to complete the new system of allotment of land not later than the 31st May, 1948, but Government expressed its anxiety to introduce the new scheme as early as might be feasible and warning was given that the cultivation of the kharif crops including sugar-cane and cotton should not be allowed to suffer. Although not in theory, in actual practice this announcement reduced the period available for work in the central office to less than a month.

During the period available claims were tabulated at much pressure. Neces-sarily, the areas claimed to have been abandoned were included in the tabulation at their face value. The results of the tabulation left a general impression that there had been a certain degree of exaggera-tion in the claims and a number who did not hold land had claimed to be landholders. On the 14th May, 1948, Government felt bound to announce that without further steps for the verification of areas said to have been abandoned, the new scheme of land allotment announced in the press communique of the 7th February, 1948, would not be launched.

Verification of Land Claims

At this stage there was no assurance that revenue records would be exchanged between the two Dominions. Correspondence on the subject had been inconclusive. There was a deep desire every-where to introduce a scheme of quasi-permanent allotment. It was decided to arrange for oral verification of claims. Verification took place at tehsils and sub-tehsils in East Punjab and East Punjab States between the 17th August, 1948, and the end of October, 1948. Centres were fixed at which the claims of individual villages of Western Pakis-tan were verified.

As a rule, a village was named against a particular centre if the largest number of claims of that village had been register-ed there. Brochures setting out the verification programmememes in detail from the aspect of West Punjab districts as well as that of individual districts in East Punjab and East Punjab States were published. Addi-tional staff was recruited for the purpose and for the best part of two and a half months the energies of the revenue administration were employed on verification. For the first time an attempt was made to draw up lists of villages abandoned in Pakistan.

Without knowledge of hadbast numbers, following in many instances wrong names given by applicants, the lists contained errors, but these were gradually reduced. A form known as the parcha tasdiq was devised for recording the results of the verification. Verifying Officers were directed to read out to persons from any West Punjab village which was under verifi-cation the names of claimants as given in the claims and to ask lambardars and other persons present if they could give information concern-ing the area, the kind of soil or the nature of rights held by each individual in their village.

After clearing discrepancies as far as possible, Verifying Officers were to record in the parcha tasdiq their conclusions about the area, the kind of soil and the nature of rights possessed by each claimant. In support of a claim, proof was to be considered in the following order of precedence:

- Entries in a jamabandi or other revenue record; or
- Entries and references in any registered document;
- Entries or references in other documents of unquestionable genuineness; and
- Oral evidence of reliable persons present at the verification.

Verifying Officers were also to record in the parcha tasdiq against each claimant whether he was personally present at the verification, whether he produced any documentary evidence, (if so. giving hrief description), and if a claimant was serving in the Defence Services, whether any relative had assisted in the verification of his claim. The assistance of non-officials was also sought in full measure. For displaced landholders residing outside East Punjab. East Punjab States or Delhi, as well as for persons serving in the Defence Forces, personal attendance at verification was not suggested. They were requested to give certain particulars and, among these, to answer the following questions:—

Do you want your allotment of land ?

- According to district, tehsil and residence of Pakistan? or
- According to tehsil or district of origin ? or
- Along with landholders of one of the villages in which land was held. If so, name the village ? or
- According to the district and tehsil of present temporary allotment; or
- According to the village of present temporary allotment ? or
- In any other manner, which may be stated.

These questions were also included in the parcha tasdiq. The pre-ferences indicated by individuals were noted and considered later. But to an extraordinary degree applications giving new preferences or preferences stated with reference to particular villages poured in when the work of allotment was actually begun. Where the object of invit-ing claims is to distribute any form of property, it is necessary to evolve In advance a suitable procedure for indication of wishes and preferences, which should then be treated as firm without repeated opportu-nity or permission to change; or to evolve a procedure which makes individual preferences largely unnecessary. The parcha tasdiq record proved on the whole a useful source of information in later work.

Exchange of Revenue Records

On the 1st July, 1948 arrangements were made between the West Punjab and the East Punjab Governments for preparation of copies of revenue records with a view to mutual exchange. Public opinion in the province, while appreciating the importance of exchange of records, was in favour of making such arrangements as would facilitate progress towards quasi-permanent allotment without being wholly dependent upon records which might be received in exchange.

The exchange of copies of revenue records between the two Govern-ments began on the 22nd November, 1948 and about two-thirds of the records due were received by the end of January. 1949. Thereafter, copies of jamabandis continued to be received at intervals in smaller numbers.

Analysis and tabulation of claims on the basis of the original Claim applications and the parcha tasdiq began in the '/curth week of November. 1948. In the first instance, as jamabandis were being received in driblets and it was not certain when the exchange would be completed, the analysis and tabulation of claims was carried out on the basis of material already available without recourse to details of individual holdings as given in jamabandis. The circumstances in which the work was done dictated the procedure which was adopted.

Under other circumstances, work on individual claims may have been deferred until all the jamabandis were available, or all the effort might have been given to the preparation of accounts from jamabandis, the claims being considered later.

It was felt, however, that from the pre-liminary work on claims and the parcha tasdiq, it was possible and necessary to obtain a picture of the problem of resettlement as a whole, so that the main lines of policy could be drawn up. The desire to com-plete the quasi-permanent allotment by the end of May, 1949 hastened work on whatever material was available. The procedure followed was, therefore, to draw material from the claims and the parcha tasdiq, check area totals for villages of each tehsil whose jamabandis had been received by about the middle of January, 1949, apply certain correction factors and arrive at approximate figures of area abandoned in each tehsil and district of West Punjab.

Thus, there was a good deal of work connected with analysis and tabulation of claims which had to be done twice over, once before claims were verified against individual jambandi entries and, on the second occasion, after such verifications had taken place. Nevertheless the size of the problem and the policy issues on which decisions of Governments were called for became apparent six months earlier than they would have otherwise done. This led to clear and timely decisions which stood '" firm later in circumstances propitious to fundamental change.

SUSTAINABLE COMMUNITY DEVELOPMENT

Dresner (2002:1) states that sustainable development is development that meets the needs of the present without compromising the ability of future generations to meet their needs. Sustainable development is multi-faceted and includes four central components namely; economic, political, social, and environmental aspects.

Sustainable economic development emphasizes the enhancing of existing local assets without degrading their quality. Political sustainability hinges upon the support of the majority of the community and must be effective over the long-term. Social sustainability requires that a project or initiative be integrated into, and connected with, a community's particular social structure. Environmental sustainability relates to the impact of a project or proposal on the local environment, with enhancement rather than degradation being the important outcome.

It is important to recognize that the concept of sustainable development with all its elements has tremendous usefulness for community development

- It introduces consideration of the long-term consequences from today's actions and decisions. This changing of the temporal framework is very different from standard economic modeling.
- It also encourages practitioners to think broadly across issues, disciplines, physical and social boundaries. It suggests searching out new ways to: a) create economic vitality, b) maintain a healthy environment, c) build healthy communities, and d) meet local needs.

Key principles of sustainable community development include:

- Fostering a commitment to place,
- Promoting community vitality,
- Building local capacity to support resilience and adaptability when confronted with change,
- Promoting a sense of responsibility as stewardship,
- Reinforcing the importance of a connection between the local and broader levels, and
- Promoting equity within the local social structure.

Local Development

Locally focused economic development (LED) offers local communities, local government, the private and not-for-profit sectors the opportunity to work together to improve a local economy. It focuses on enhancing competitiveness, increasing sustainable growth and, in some cases, ensuring that growth is inclusive.

Traditional LED links a range of disciplines including physical planning, economics and marketing. It also incorporates many local government and private sector functions including environmental planning, business development, infrastructure provision, real estate development and finance.

To achieve this, localities are required to:

- Develop a holistic strategy aimed at growing local firms
- Provide a competitive local investment climate
- Support and encourage networking and collaboration
- Encourage the development of business clusters
- Encourage workforce development and education
- Closely target inward investment to support cluster growth, and
- Support quality of life improvements.

Local communities respond to their LED needs in many ways, and a variety of approaches can be taken that include:

- Ensuring that the local investment climate is functional for local businesses;
- Supporting small and medium sized enterprises;
- Encouraging the formation of new enterprises;
- Attracting external investment (nationally and internationally);
- Investing in physical (hard) infrastructure;
- Investing in soft infrastructure (educational and workforce development, institutional support systems and regulatory issues);
- Supporting the growth of particular clusters of businesses;
- Targeting particular parts of the city for regeneration or growth (areas based initiatives);
- Supporting informal and newly emerging businesses;
- Targeting certain disadvantaged groups.

By the year 2025, close to half the poor around the world will live in urban areas. Unprecedented urbanization in the third world requires a sustainable development strategy to improve the quality of urban management and foster an economically competitive environment. Without a city development strategy there will be decrease in welfare and quality of life for urban inhabitants. However, the high concentration of persons in cities suggests that with the proper approach to growth the benefits of development can be more widely dispersed.

The World Bank not only works with leaders on the international and national level but on the local level as well. Since greater decentralization in recent years increased the responsibility of municipal or local government, it is the aim of the Local Economic Development (LED) Specialists in the

Urban Development Sector to assist local governments in determining the most effective strategy to increase jobs and revitalize their city's economy. Whereas urban development in the past included primarily critical but non-holistic approaches, such as infrastructure improvement, the new urban agenda is now deepening to encompass a sustainable and self-regenerating approach that corrects for market failure. The LED approach is effective not only in urban areas but also in peri-urban and rural areas. The LED strategic approach assists local governments in pursuing good practices in building environments that are livable, competitive, well-governed and managed, and bankable.

The LED approach acknowledges autonomy of the local government. Thus, the objectives of LED initiatives are to encourage local participation and consensus building in determining economic and social welfare initiatives for the community. While focusing the local economy and the importance of local ownership of the development process, the strategic LED approach concurrently views development within the context of the governance and civil society on all levels. The LED strategy reflects the view that urban areas can effectively contribute to the national government through public policies coupled with community action, private sector commitment, accountable local government, and supportive national government.

The contemporary notion of development was born in New Hampshire at the Bretton Woods Economic Summit. The Summit which was organized at the end of World War II resulted in the establishment of two of the most powerful financial institutions of the 20th Century; the International Monetary Fund, and the World Bank. Initially designed to finance the reconstruction of Europe after the far, they functioned in an environment sold to the benefits of capitalism. Owing to fifty years of intervention in less developed countries, and implementation of "structural adjustment programs" and "poverty reduction papers", famine, poverty, environmental disasters, and food and energy crisis constitute the lot of very many less developed countries around the world. No wonder a great majority of economists are now demanding the reform of these institutions. The models advocated by the IMF and the World Bank simply did not work.

In traditional development, bigger is considered better, with large-scale, centralized projects which necessitate substantial infusion of capital. Increased Gross National Product, per capita income, and worker's productivity all constitute the quantitative means by which development in measured. Usually, development initiatives conceived and designed using quantitative criteria expose sets of statistical data which say little about the impact of those initiatives on the lives of the people. Many economists, might we say, with a "social conscience" began to introduce other disciplines into economics with a focus on quality of life. In the mid-1970s, the concept of Physical Quality of Life Index was developed by Morris David Morris in response to the World Bank and the International Monetary Fund policies of growth.

For the calculation of PQLI, Morris came up with three indicators; life expectancy at age 1, infant mortality, and literacy. Though his study found that countries with low per capita incomes were more likely to have low quality of life, and country with high per capita incomes tended to have high quality of life, he could not establish close correlations between GNP and PQLI.

With his study of a number of countries, "developed" and "underdeveloped", he set out to demonstrate flaws in the use of the gross national product (GNP) as a main indicator of the level of development of a country. Morris then shifted the focus from GNP to the quality of life instead. He discovered that some countries with a high GNP did not rank high on the PQLI, while some of the countries with very low incomes scored very high on the quality index.

Most economists recognize that it is hard to quantify "quality of life". They also agree that the foremost purpose of development must be to provide people with adequate food and shelter, economic security, and equitable distribution of resources, all elements forming the base of social capital.

The new attitude and thinking on development was in line with the first 1960-1970 United Nations Decade was driven by the notion that development meant growth plus change in all aspects including social, cultural, qualitative and quantitative. The center piece was quality in people's life.

Following the same concerns, the 1974 Declaration of Cocoyoc in Mexico emphasized the "development of man not of things" and stressed the need for diversity and the right to follow different roads to development so as to achieve self-reliance. Similarly, in 1975, there was a call for "another development" focusing on human-centered development. In 1977, UNESCO in Paris suggested a total multi-relational process that includes all aspects of life of a collectivity, of its relationship with the outside world and of its own consciousness. The Brundtland Commission has recently recommended sustainable development for our common future.

Poor nations, after decolonization, were encouraged to adopt the western model of modernization as a way out of abject poverty. Some former colonies tried their hands at the "basic needs" approach and the participatory approach with very limited success and a good dose of confusion. In the midst of all this, there is an ongoing debate between neo-liberal economists and globalization supporters. This debate is fiercely challenged by the centre-left, post-capitalist Third Way, and by Another Way is Possible of the anti neo-liberal World Social Forum.

Today, most economists contend that more economic growth is not synonymous of development. Equally, it is clearly understood that development and human development do not necessarily translate into well-being and happiness.

Recent development initiatives tend focus on people, communities, grass-roots, redistribution of wealth, social capital, environment, culture, tradition, local or indigenous knowledge, participation, diversity, local ownership, sustainability, interdependence, multidisciplinary approaches, connectedness, and more.

There is a strong movement and commitment from a group of development practitioners to redefine development in terms of well-being, human development, sustainability, human rights and happiness. Consequently, traditional development indicators like gross national product purely based on economic criteria are increasingly deemed insufficient and seriously challenged. In the 1970s, the term "Gross National Happiness" was coined by the 4th King of Bhutan as a substitute to "Gross National Product" or "Gross Domestic Product". His move has inspired a growing international movement who are revitalizing the directions of "progress". In February 2004, his son, the 5th King of Bhutan, H.M. Jigme Khesar Wangchuk stated at the First International Conference on Gross National Happiness: "I feel that there must be some convergence among nations on the idea of what the primary objective of development and progress should be – something Gross National Happiness seeks to bring about".

At the Third International Gross National Happiness meeting in Bangkok held in November 2007, the outgoing Prime Minister of Thailand, Surayud Chulanont affirmed:" An enormous challenge is to develop new ways to measure appropriately the success or failure of innovative policies. Using old standards for new policies will not help. I do hope that this Conference can help to advance our search for new approaches and indicators that will enhance contribution to sustainability. A cooperative framework for research should not only include Thailand and Bhutan, but also our colleagues within ASEAN and indeed from all regions of the world".

The Right to Development, which was proposed in 1972 by the Senegalese jurist Keba M'baye, remains controversial to this day. This concept was given legal recognition in the 1981 African Charter on Human and People's Right, to be later incorporated in the Global Human Right Framework through the adoption in 1986 of the Declaration on the Right to Development by the

United Nations General Assembly. The 1993 Vienna Declaration and Programme of Action, the 2000 Millennium Declaration, and the recent Durban Declaration and Programme of Action reaffirmed the Right to Development as a universal and inalienable human right. A number of elements constitute the core content of the Right to Development: comprehensive development, respect for all human rights, participation, social justice, international cooperation, and self-determination.

Even though the content of the Right to Development is straightforward, clarifying the nature of the right is more challenging. Added to the fold as a series of concerns like: what kind of right is it? Who are the duty-holders and right-bearers? How can it be implemented, monitored, and enforced? Better yet, there is a lack of consensus regarding the meaning and status of the Right to Development, provoking a fierce political debate among academicians, northern governments, southern governments, international organizations, and non-governmental organizations.

As in the search for consensus on the definitions and contents of new development approaches, agreeing on sets of indicators to comprehensively reflect new definitions of development present daunting challenges. However, a survey of development indicators used the past decades suggests a strong tendency to incorporate new indicators that are not necessarily easy to measure.

The Human Development approach, which is interested in building capabilities and the abilities to function, concentrates on knowledge, a decent standard of income, and a long and healthy life. Elements of human development are productivity, sustainability, equality and empowerment. Human Development Index can be listed as: 1) life expectancy, 2) education measured by combined estimate of adult literacy and average years of schooling, and 3) economic standard of living in terms of purchasing power after adjusting for the local cost of living.

Developed by Hazel Henderson, the Country Futures Indicator is a more comprehensive approach with a bias in cross-cultural characteristics. These indicators suggest criteria to make governments and institutions more accountable using the principles of social, democratic, and environmental audit. Some of the indicators are: 1) income distribution, 2) informal and household sector production, 3) deduction of social and environmental costs, 4) community-based accounting, and 5) military / civilian budget ratio. In addition, some complementary indicators of progress include: 1) education, 2) health, 3) nutrition, 4) basic services, 5) shelter, 6) political participation and democratic progress, and 7) status of minorities, women, etc. In the United States, the project on Redefining Progress has developed the Genuine Progress

Indicators. They take into account quality and distribution of economic growth. Principally considered are: 1) value of housework, 2) care for children and the elderly, 3) volunteer work, and 4) free time or family and community activities. Even though money is not exchanged, these activities contribute to the economy and the well-being of people. In Canada, indicators developed include democracy index. They have suggested projects to include ecological footprint, common assets, climate justice coalition and environmental tax reform.

UNESCO and the United Nations Research in Social Development have proposed cultural indicators, such as: 1) the Cultural Freedom Index which includes freedom of thought, belief and expression, 2) the Creative Empowerment Index which includes encouragement to innovative expression, and 3) the Cultural Dialogue Index which includes opportunities to mutually benefit communication among people of different cultures.

A Welfare Index in observance in Scandinavian countries combines the three aspects of having, loving and being. The index includes: 1) income, 2) quality of housing, 3) political support, 4) social relations, 5) health, 6) education, 7) being irreplaceable, 8) doing interesting things, and 9) life satisfaction.

The World Health Organization Quality of Life indicators include: physical environment, home environment, financial resources, social support, safety, information, and transportation. The inner quality consists of physical health, mental health, work capacity, learning capacity, energy, absence or presence of pain and depression, satisfaction with self, and satisfaction with life.

The Council for Social Development in Canada describes quality of life in terms of being, belonging and becoming. Being includes physical health, psychological thoughts and feelings, and spiritual beliefs and values. Belonging includes physical as in a living place, social as in having people around, and community resources. Becoming includes a practical as in daily things to do, leisure for fun and enjoyment, and growth as in coping with change.

A much simpler approach is advocated by the WIDER Institute in Finland. This approach consists of outer qualities, which means living in a good environment and being worth for the world, and inner qualities, which means being able to cope with life and enjoy life.

DECENTRALISATION: CONCEPTS AND THEORIES

Terms and Concepts

Decentralisation can be usefully understood as a political process whereby administrative authority, public resources and responsibilities are transferred from central government agencies to lower-level organs of government or to non-governmental bodies, such as community-based organizations, 'third party' non-governmental organisations or private sector actors.

Conceptually, important distinctions can be made among:

- Deconcentration, in which political, administrative and fiscal responsibilities are transferred to lower units within central line ministries or agencies;
- Devolution, in which sub-national units of government are either created or strengthened in terms of political, administrative and fiscal power;
- Delegation, in which responsibilities are transferred to organisations that are 'outside the regular bureaucratic structure and are only indirectly controlled by the central government,';
- Privatisation, in which all responsibility for government functions is transferred to nongovernmental organisations or private enterprises independent of government.

Such transfers can involve the power to decide the allocation and distribution of public resources, the power to implement programmes and policies and the power to raise and spend public revenues for these and other purposes. These three powers we can classify broadly as political, administrative and fiscal decentralization. However, democratic decentralisation implies more than the downward delegation of authority.

Crucially, it entails a system of governance in which citizens possess the right to hold local public officials to account through the use of elections, grievance meetings and other democratic means.

Blair captures the essence of this important idea:

- Can be defined as meaningful authority devolved to local units of governance that are accessible and accountable to the local citizenry, who enjoy full political rights and liberty. It thus differs from the vast majority of earlier efforts at decentralization in developing areas, which go back to the 1950s, and which were largely initiatives in public administration without any serious democratic component.

A defining feature of any democratic system is that decision-makers are under the 'effective popular control' of the people they are meant to govern. How this is accomplished, of course, constitutes a major dilemma for theorists and proponents of democratic development. Nevertheless, a number of defining features can be observed.

Mayo identifies four:

1. Popular control of policy makers, both by regular elections and by the pressure of social interest groups;

2. The institutionalisation of all adult citizens in voting;
3. Political freedom in the eyes of the state;
4. Policy decisions made on the basis of majority rule.

Beyond these very basic principles, democracy also implies a wide range of rules, norms and customs through which citizens can exercise 'effective popular control' over public officials. Included here would be an independent judiciary, a free press, systems of transparency, and freedom of association and speech. We are particularly interested in the intra-governmental transfers that fall within the spectrum of deconcentration and devolution, rather than extra-governmental transfers, such as delegation and privatisation. Panchayati Raj in India is primarily concerned with devolution, in which the 73rd Amendment confers constitutional status to a 'third stratum' that exists within the government system at district, sub-district and village levels. For the sake of brevity, we assess fiscal, political and administrative decentralisation in AP and MP simply in terms of the extent to which the State governments have devolved powers over funds, functions and functionaries.

The Case for Decentralization

Assertions in favour of decentralisation are often founded upon a wider critique of central state planning, which holds that large and centrally-administered bureaucracies represent an inefficient and potentially destructive means of allocating resources within society. Three assertions are used to substantiate this claim. One argues that central state agencies lack the 'time and place knowledge' to implement policies and programmes that reflect people's 'real' needs and preferences. Decentralisation is thought to create the conditions for a more pluralist political arrangement, in which competing groups can voice and institutionalise their interests in local democratic forums.

A second holds that states are qualitatively different from markets and voluntary organisations. Viewed in this way, states lack the flexibility and reach to provide certain types of goods and services, particularly ones with large information requirements. Decentralisation, it is argued, creates institutions that are more amenable to local needs and preferences. A third and related view argues that unchecked authority and inadequate incentives encourage 'rent seeking behaviour' among government officials.

In theory, decentralisation would undermine these opportunities by creating institutional arrangements that formalise the relationship between citizens and public servants, giving the former the authority to impose sanctions on the latter. This document is particularly concerned with this third aspect of decentralisation.

Decentralisation and Poverty Reduction

The case for democratic decentralisation is also predicated upon the notion that greater participation in local political affairs will improve the quality and reach of government services, particularly ones aimed at improving the lives of poor and politically marginal groups in society. For proponents of democratic decentralisation, a central challenge of improving the delivery of public services becomes one of 'crafting' institutions which can maximise participation in political life. In the context of poverty reduction, access to the resources and benefits that governments provide is associated with systems of governance that empower poor and vulnerable groups in society. lthough there is of course great variation among cultures, countries and regions, we can identify a number of roles that governments typically play in poor and predominantly rural areas:

- One is the provision of public goods, such as universal education and healthcare;
- A second is the provision of divisible goods, such as irrigation, agricultural extension and credit;
- A third is the determination and enforcement of laws regulating key economic inputs, such as land, labour and capital;
- A fourth and critical element is the recognition and protection of rights allowing for organisation, association and entitlement in the eyes of the state.

All of these, it is worth emphasising, are problematic in the sense that they require systems of governance which ensure that public resources are being delivered efficiently and effectively. Indeed, it is the misallocation or 'corruption' of these services that often justifies the strongest calls for public sector reform. Studies of decentralisation have shown that devolution of authority can enhance systems of local governance in a number of ways. First, the establishment and empowerment of local resource user groups can improve the ways in which local people manage and use natural resources, thereby improving the resource base on which poor people are often disproportionately dependent.

Such arguments are generally made in relation to the provision of local public goods, such as common pool resources or local credit organisations. Second, and related to this, collaboration between public agencies and local resource users can produce 'synergistic' outcomes, in which citizens and civil servants cooperate to provide goods that would be unobtainable were they acting alone. Classic examples of this would include joint forest management, fisheries co-management and participatory watershed management. Third, and most central to this document, the democratisation and empowerment of local administrative bodies can enhance participation in decision-making fora, particularly among groups that have traditionally been marginalised by local political processes.

Studies from Africa, Asia and Latin America have shown that the introduction of elections, systems of transparency and rights of association can empower poor people, enhancing their ability to participate in local decision making and encouraging them to hold public officials to account. As Blair points out, 'increased representation offers significant benefits in itself.' Framed in this way, participation in local, democratically elected bodies can lead to improvements in self-identity and worth, which can help to break down customs of inequality and discrimination. Second, membership of local administrative bodies can provide important skills that can be transferred to other walks of life. However, the notion that improving participation through decentralisation will necessarily lead to improvements in people's wellbeing is not entirely consistent with documented evidence.

A recurring theme that emerges from a sizeable body of literature is the relatively weak correlation that exists between democratic decentralisation and poverty reduction. Despite great strides at devolving power to local, democratically elected bodies, decentralisation in Colombia, Brazil and the Philippines appears to have achieved little in the way of reducing poverty or improving regional disparities. Manor's conclusions about experiences in Bolivia, Karnataka and Bangladesh are equally pessimistic.

In one respect, this reflects the difficulty of establishing a clear and rigorous link between changes in governance and improvements in well-being. It also suggests a more sobre conclusion about the ways in which democratic decentralisation is thought to affect the quality and availability of local economic opportunity.

As Crook and Sverrisson's cross-country comparison concludes:

- The notion that there is a predictable or general link between decentralisation of government and the development of more 'pro-poor' policies or poverty-alleviating outcomes clearly lacks any convincing evidence. Those who advocate decentralisation on these grounds, at least, should be more cautious, which is not to say that there are not other important benefits, particularly in the field of participation and empowerment.

Even the most successful forms of democratic decentralisation have been unable to overcome economic and political disparities, both within and among regions. This, in part, highlights the problem of raising public revenue in rural areas, in which economic surplus is typically poor. Decentralisation can also pose new problems of co-ordination and planning. As the Kerala experience has shown too much devolution can lead to major duplications of effort and gaps among different government agencies. Without adequate training and support, the devolution of large sums of money can also over-burden local bodies whose members lack the resources and expertise to spend large and complex budgets.

Finally, there is a problem of local élite capture. As many studies have pointed out, one of the dangers of devolving authority is that it simply empowers local élites and, worse, perpetuates existing poverty and inequality. Whether the introduction of democratic principles–on its own–would overcome the historical and cultural factors that perpetuate political inequality is somewhat doubtful. This, in turn, highlights the challenge of encouraging democracy in rural areas in which large numbers of people are dependent upon small numbers of local, powerful élites. It also highlights the challenge of encouraging 'empowerment' without addressing the rights and entitlements that underlie political structures in rural areas. Where the spoils of government intervention are particularly good, one can predict with reasonable confidence that the costs of ensuring equitable distribution and of discouraging local corruption will be high. This is notoriously true of rural infrastructure projects, such as road building or irrigation in which markets for primary inputs, labour and public regulation are strong.

Here the corruption of local administrative bodies will depend on a number of factors:

- The ease of procuring spoils from the programme;
- The ability to avoid detection and/or sanction;
- The amount of money involved;
- Systems of accountability, which would ideally expose and sanction behaviour of this kind.

This last factor is worth highlighting, both because it can be affected by public policy and because it is the misallocation or 'corruption' of public resources that often justifies the strongest calls for democratic decentralisation.

Saxena and Farrington highlight a number of ways in which poor accountability and bad governance can undermine the interests of poor and marginal groups in society:

- First, rent seeking behaviour on the part of the police, government officials, etc. Can deprive the poor of resources they would ideally receive were the corruption not taking place;
- Second, absenteeism, corruption and other forms of poor performance

means that the government is spending social welfare resources which yield no benefits;

- Third, the widespread perception of corruption promotes feelings of distrust towards government, thereby undermining the potential for collaboration between the state and civil society organisations;
- Finally, a culture of corruption emerges within the public sector, encouraging officials further rent seeking behaviour and poor accountability.

Perhaps the strongest argument in favour of decentralisation is the idea that the establishment of strong and accountable systems of governance will encourage a more equitable distribution of stateprovided resources, which can be directed towards poor and vulnerable groups. An underlying hypothesis here is that democratic participation will yield strong mechanisms of accountability, which in turn, will improve the distribution of benefits to groups that are traditionally marginalised by market and state. We explore this hypothesis more fully by examining the conditions under which decentralisation has led to more accountable forms of local governance.

The conclusion we reach is that decentralisation–on its own–is far from sufficient. First, however, it is important that we try to define more precisely what we mean by accountability. Agrawal and Ribot have argued that accountability implies a set of relations that depend 'on the exercise of a counter power to balance arbitrary action.' Comparing decentralisation in South Asia and West Africa, they draw a useful distinction between upward and downward accountability, in which the ability to hold decision making actors to account is dependent on the upward and downward relations between officials and their superiors and local citizens.

Agrawal and Ribot argue that the most effective forms of governance are dependent on downward accountability, in which the power to make rules, decide access to the benefits provided by local bodies and enforce and adjudicate these arrangements is dependent upon the consent or support of local citizens. This, in turn, implies that local people have the power to hold officials in check. Brett's definition of accountability implies a more formalised set of expectations and responsibilities, on which the performance of public officials can be judged. Here the existence of clear and explicit rules, governing the duties, jurisdictions and standards of public officials is an essential part of accountability. As he argues in later work, 'Accountability works best when rewards depend directly on the quality of service provided–failures occur when there is no direct relationship between the two.' Moncrieffe distinguishes between ex-post and ex-ante accountability.

The former, she argues, implies a responsive process on the part of citizens, whereby rights-based mechanisms, such as the courts, grievance procedures and elections, can be used to keep public officials in check.

The latter suggests that:

- ... in order to act effectively in the citizens' interest, representatives must–as a general principle–know what these interests are; allow for deliberation and consultation so that policies may be corrected where appropriate; keep the public apprised of policy choices and provide explanations and opportunities for public response, particularly where it is not obvious that actions are in accordance with the public's expressed interests; and provide the appropriate mechanisms and act in such a way that citizens are able to assess the quality of their representation.

This definition–and the relations it implies–comes closer to Judith Tendler's interpretation of good governance, a theme we turn to next. To summarise, the most ambitious forms of decentralisation constitute a substantive shift in power from national or regional levels to more local spheres of political life.

Decentralisation empowers new actors and creates conditions for new lines of participation and accountability. However, the notion that decentralisation will lead to improvements in poverty reduction is not well-supported. Moreover, there is little or no evidence to suggest that decentralisation–on its own–will necessarily produce systems of governance that are more effective or accountable to local needs and interests.

In other words, local forms of accountability are dependent on more than just the devolution of political, administrative and fiscal power. We consider arguments and evidence outlining the conditions under which local institutions can be made more accountable. In the context of this document we define accountability as a relationship between public officials and citizens, in which the latter possess a means of challenging or counterbalancing the arbitrary use of power on the basis of a formal understanding of what constitutes appropriate behaviour on the part of public officials. In its most ideal form, accountability should be based on strong norms of communication and consultation between public officials and citizens.

The 'Role' of Civil Society

Central to much thinking about governance and accountability is the notion that effective and responsive governments require strong and vibrant civil societies to keep them in check. Perhaps the most recent and influential manifestation of this is Robert Putnam's assertion that societies with high levels of social capital will organise to demand better government. Underlying this proposition is the notion that 'civic engagement'–participation in a wide range of political and non-political organisations–correlates strongly with effective and responsive government.5 Civil society is often understood as a 'sphere' of voluntary action, which sits between the family and the state.

Included here would be 'third-party' organisations, membership organisations, political parties, Mafia arrangements, religious affiliations and firms. The value of defining civil society in this way is that it attempts to differentiate between groups and relations that are organised on principles of hierarchy and control and ones that are based on the relatively altruistic motivations of the nuclear and extended family unit. The drawback, of course, is that it underplays the normative and ideological terms on which people join and establish social groups in the first place. Studies of decentralisation suggest two important links between civil society organisations and local accountability.

One emphasises the importance of local mobilisation. Drawing on evidence from China and Nigeria, Meenakshisundaram argues that local accountability was highly dependent on the existence of 'mass participation' at the local level. Reflecting on findings from Bolivia, Honduras, India, Mali, Ukraine and the Philippines, Blair argues that civil society organisations fostered strong accountability in peripheral rural areas.

An important assertion here is that accountability is strengthened when people outside the state organise to ensure that public officials are acting in accordance with formal and informal norms of performance. A second assertion is that external civil society organisations can empower poor and marginal groups in society. NGOs, for instance, have been shown to empower poor people by connecting them with a wider circle of allies, with whom they can mount a more effective political lobby. Second, and related to this, they can absorb some of the costs of engaging in political action. Third, and somewhat less tangibly, they can encourage what Samuel Popkin has described as 'new conceptions of identity and self-worth.' This they can do by encouraging poor people to engage in collective action or by transmitting information about constitutional rights, potential allies and other political opportunities.

However, it is important to recognise the fact that civil society organisations are often constrained by institutions that counteract or undermine local forms of political action. In rural communities, where the number and diversity of social networks are often limited, individuals obtain access to important material resources by claiming, maintaining and ideally improving their place in lineage groups, such as the family, and membership groups, such as religion and 'community'. To assume that these groups would be open to all that seek to join would understate the benefits insiders can obtain by restricting membership. A direct illustration of this is the relationship between literacy and political action. As Drèze and Sen have argued, one's ability to obtain and understand information about laws, policies and the rights to which one is entitled is often highly dependent on the ability to read.

This, in turn, highlights the means by which poor people are represented in democratic institutions and the extent to which they have the 'political tools' to influence the democratic process. When voters are ill informed about party platforms, government policies and the rights that these may provide, their ability to influence the democratic process can be limited. Likewise, when politicians and parties campaign on the basis of short-term pay-offs, as opposed to programmatic policies, the relationship between compromises of this nature and democratic accountability can be very thin indeed. Related to this is the dilemma of encouraging poor people to assume the costs of engaging in direct political action.

As Moore and Putzel have argued, agrarian institutions may be structured in a way that prevents poor people from engaging in direct political action. Moreover, the costs of political action may deter them from pursuing or sustaining coherent political movements. Finally, multiple and potentially contradictory loyalties may undermine political solidarity around class-based identities, such as 'small farmers, landless, wage workers, tenants, recipients of food subsidies, squatters' and the like.

Reflecting on the recent popularisation of civil society, social capital and decentralisation, Harriss has argued that the use of social capital and civil society as the basis for the establishment of good governance is a dangerous assertion in the sense that it implies that local collective action is sufficient for strong mechanisms of accountability.

In particular, he argues that local organisation and collective action on their own are 'liable to be ineffectual' without the active support of external civil society organisations and the bureaucratic state.

Emphasising the notion that civil society is established 'in relation to institutions which are defined by the state,' Harriss argues that collective action will improve local governance structures only when they are connected to wider networks of political action and established in 'a political context which secures the rights of less advantaged or less 'resource-full' people.' In other words, local people require the support of higher authorities within the state, whose power can support and legitimise local forms of political mobilisation.

Regulations stipulating the inclusion of subordinate groups can help to ensure that poor and marginalised groups have a voice in local bodies. Among the most successful cases, systems of local governance have also been shown to improve the efficiency and responsiveness of public officials. However, without these forms of support, poor and primarily rural communities are unlikely to join and sustain political movements on their own. This is consistent with a wider literature on decentralisation, which emphasises the importance of 'enabling regimes' that allow for local autonomy.

'Enabling Regimes' and Local Autonomy

Judith Tendler's study of governance and primary healthcare in Northeastern Brazil helps to show the ways in which extended interaction between government health workers and local communities was able to foster a culture of accountability between public officials and the rural poor. Central to Tendler's analysis is the 'paradoxical' notion that the effective delivery of primary healthcare was dependent on external support from 'higher-level' echelons within government.

This is consistent with a wider literature on decentr-alisation, which suggests that central governments can foster local accountability in a number of ways. First, as Crook and Sverrisson have argued, central states can provide an important 'counter élite' to groups that would resist efforts to make local bodies more democratic. Second, and crucially, they can structure incentives in a way that allows local participation and public accountability to take root. Such incentives would conceivably include career trajectories, 'earmarked funding' for local bodies and status within society.

Drawing on evidence from China and Nigeria, Meenakshi-sundaram argues that effective decentralisation was dependent on the existence of three necessary conditions:

1. Strong political commitment from higher level authorities within government;
2. Relative autonomy of the local body in decision making and implementation of local schemes;
3. The availability of internally generated resources at the local level.

These are broadly consistent with the conclusions reached by Crook and Manor in which:

- Elected bodies at local levels had adequate funds;
- They enjoyed substantive autonomy; and
- Lines of accountability existed between elected representatives and citizens, and between nonelected bureaucrats and elected representatives.

In other words, devolution can foster accountablity when local institutions have the autonomy to decide matters relating to local revenues and resource allocation. However, where the international literature is somewhat more ambiguous is on the question of how to strike a balance between the autonomy that local bodies need to function effectively and the accountability to ensure that such bodies act in the public interest. Evans addresses this issue when he considers the question of state autonomy and industrial policy in Brazil.

His principal assertion is that effective state intervention required a 'concrete set of social ties that binds the state to society and provides institutionalised channels for the continual negotiation and renegotiation of goals and policies,'. At the same time, he asserts, government officials required sufficient institutional autonomy to withstand the influence of powerful actors and interests. This, again, is consistent with Moncrieffe's idea of 'ex-post' accountability.

Along similar lines, Tendler's study found that accountability and good governance were contingent upon the following conditions:

- Government officials spent extended periods of time with beneficiaries;
- This, in turn, created a situation in which officials were affected by the opinions and sanctions of community members;
- Good performance carried high prestige, both within the community and within the civil service;
- Central government was instrumental in supporting these initiatives.

In theory, local taxation and the threat of the vote would strengthen the autonomy and accountability of local bodies.

As Jha has argued:

- The logic behind transferring taxes which are local in nature to the local bodies is to endow these bodies with strong revenue base that can be utilised according to their own development plans.

In practice, however, the decentralisation of taxation may be a difficult undertaking. As Manor has argued, the central dilemma here is not necessarily the lack of taxable surplus but the political and administrative costs of collecting public resources, the reluctance among many central governments to grant the authority that activities of this nature would require and the reluctance among residents to in fact pay their taxes. Lacking a means of ensuring transparent budget allocation, there is also little guarantee that the creation of financially autonomous bodies will not simply perpetuate further corruption of local resources or, worse, encourage local politicians to eliminate certain services entirely. Such findings are highly consistent with the problems most commonly associated with decentralisation in India.

Elections, Parties and Competitive Politics

Crook and Manor's study of decentralisation in South Asia and West Africa highlights the challenge of encouraging a culture of accountability in local political processes.

Reflecting on the relatively successful case of Karnataka, they argue that accountability required the existence of:

- Competitive political parties;
- A widely distributed free press; and
- A 'professional civil service,' in which officials were willing 'to develop a constructive but lawabiding relationship with elected politicians'.

Reflecting on the findings of a USAID study of democratic decentralisation, Blair identifies seven mechanisms 'that have proven at least potentially viable' in Bolivia, Honduras, India, Mali, Ukraine and the Philippines:

- Free and fair local elections;
- Strong and competitive political parties;
- Strong civil society organisations;
- A vigourous and accessible electronic media;
- Public meetings;
- Formal grievance procedures.

As Blair has argued, periodic elections provide an important means of ensuring government responsiveness and accountability on broad social issues. At the same time, he observes, 'elections are crude instruments of popular control, since they occur at widely spaced intervals... and address only the broadest issues,'. Elections therefore constitute an imperfect yet vital component of any democratic system. However, their ability to encourage effective responsive governance is highly dependent upon three important variables:

- The degree to which parties and politicians campaign on substantive policy issues, as opposed to populism or, worse, clientelism and vote buying;
- The quality of information voters have at their disposal; and
- The strength of civil society organisations.

As Manor has argued, multi-party democracies can help to organise 'opposing forces' into clearly recognisable groups, stimulating public criticism and debate. That having been said, the ability to articulate interests and stimulate debate depends in no small way on the internal dynamics and debates that exist within political parties. Just as elections do not a democracy make, the same can be said of party politics and democratisation. The development of a strong and vibrant civil society is also inextricably linked to the political opportunities the state makes available, and the ways in which poor and marginal groups in society exploit these opportunities. This, in turn, highlights the ways in which identities based on class, caste, religion, ethnicity, gender and other social markers affect social mobilisation and political voice.

It also poses the question of whether and to what extent political struggles among and between these and other social groupings will lead to stronger forms of civil society and, by extension, more equitable forms of governance. The experience in West Bengal suggests that political parties can and will challenge the interests of dominant groups when they develop and pursue a programme that is ideologically committed to the goal of social redistribution.

Kohli makes the case that the Left Front government's ability to pursue and implement a pro-poor agenda of land reform and decentralisation was directly dependent on the structure and ideological commitment of the CPI. In particular, Kohli argues that the Left Front's ability to penetrate the countryside and to challenge the interests of landed élites was dependent on a party with:

- A coherent leadership;
- An ideological and organisational commitment to exclude propertied interests from the process of governance;
- A pragmatic attitude towards facilitating a non-threatening environment for propertied interests; and
- An organisational structure that was both centralised and decentralised, allowing the regime to maintain contact with local society, without becoming beholden to local propertied élites.

Much like Tendler's 'optimal' arrangement in Brazil, the Left Front government appears to have been able to strike an ideal balance between local governance and a central executive, whose power and legitimacy helped to maintain a minimal sphere of autonomy from local élite capture.

However, the historical events that led to the establishment of the Left Front government in West Bengal have prompted some scholars to question the viability of replicating the experience in other political settings. Moreover, it is worth emphasising that the achievement of this political programme was not entirely democratic in character, reiterating the tension that can exist between coherent policy and popular democracy.

As Corbridge and Harriss have argued, 'West Bengal is not a 'model' for the rest of the country.' Nevertheless, they emphasise, it does shed some light on the important ways in which the state can empower poor and politically marginal groups in society.

DECENTRALISATION IN INDIA

A commitment to the reduction of poverty has been a defining characteristic of the Indian state, from the time of Independence to the present day. As Kohli has argued, the Indian state that emerged after Independence was deeply committed to 'industrialisation, economic growth and a modicum of income redistribution.' In terms of poverty reduction, this involved an early attempt at improving agricultural productivity through the implementation of land reforms, agricultural cooperatives and local self-government. From an early stage in this process, the reduction of poverty and the empowerment of poor and politically marginal groups in India have been strongly associated with at least some form of decentralisation.

Perhaps the most enduring image of decentralisation in India is Gandhi's vision of village Swaraj, in which universal education, economic self-sufficiency and village democracy would take the place of caste, untouchability and other forms of rural exploitation.

Although this vision has been hotly debated since the time of independence, Gandhi's vision has had an enduring effect on the ways in which decentralisation has been argued and defended in Indian politics.

Beyond the symbolic imagery of the independent 'village republic,' an important element of this relates to the idea that formal, constitutional changes in India's administrative system can have a lasting impact on informal and unequal structures like caste, class and gender.

The various commissions and committees that have inspired contemporary thinking about Panchayati Raj in India. Perhaps the most important among these–particularly since independence–were the B. Metha Commission of 1957, the Asoka Metha Commission of 1978, and the G.V.K. Rao Committee of 1985.

An enduring issue that features in all of these assessments is the notion that the Panchayats have been weakened or undermined on three fronts:

- States that are unwilling to devolve substantive power;
- A resistant bureaucracy and
- The power of 'local élites.' Such realisations were instrumental in the drive to give the Panchayats constitutional status in the 73rd Amendment.

The 73rd Amendment

The 73rd Amendment gives village, block and district level bodies a constitutional status under Indian law.

The more important features of the Amendment are:

- To review the annual statement of accounts;
- To review reports of the preceding financial year;
- To review and submit views on development programmes for the following year;
- To participate in the identification of beneficiaries for some government schemes.

This last provision is particularly important because it confers substantive authority over an area that is particularly prone to misallocation and corruption.

As Jha has argued,

- The 73rd and 74th Amendments... are designed to promote self-governance through statutory recognition of local bodies. The latter are expected to move away from their traditional role of simply executing the programmes handed down to them by higher levels of government and to formulate and implement their own programmes of economic development and social justice.

The 'Eleventh Schedule' of the 73rd Amendment identifies 29 areas over which Panchayats can legitimately have jurisdiction.

Many of these–such as agriculture, minor irrigation, animal husbandry, fisheries, social forestry, small-scale industries, and implementation of land reforms–focus on particular sectors within the rural economy.

Others–such as rural housing, rural electrification, transportation and communication linkages–are primarily concerned with the provision and maintenance of rural infrastructure.

Some cover the provision of key rural services, such as health, sanitation and primary, secondary and vocational education. Others still govern the provision of targeted welfare benefits–such as the PDS, and benefits for scheduled castes and tribes, women and children and the handicapped. In short, the 73rd Amendment covers many areas that would enable the Panchayats to improve the lives and wellbeing of poor and vulnerable groups. Moreover, it contains specific provisions that guarantee the participation of traditionally excluded groups, such as women, SCs and STs, and transparency for local institutions such as the Gram Panchayats and the Gram Sabha.

However, as Mukarji has argued, the ambiguity surrounding the concept of 'self-government' and the substantial power that still rests with the State governments have prevented most States from devolving any substantive power to the Panchayats. Such findings are consistent with a wider literature on the problems most commonly associated with decentralisation in India.

In order to consider these hypotheses in relation to India's decentralisation process, it is first necessary to establish whether the constitutional changes that transpired in 1993 were in fact decentralisation. The scholarly literature on Panchayati Raj tends to move in two fairly incongruous directions. One, which is firmly grounded in the liberal democratic paradigm, argues that the 73rd Amendment did indeed create the conditions for decentralisation in India, and that this was a positive development.

Although there is some disagreement about whether and to what extent the reforms have been implemented, the general consensus within this school is that the 73rd Amendment was a positive step for Indian decentralisation and that the only factor preventing successful devolution is the failure to implement fully the terms and conditions of the 73rd Amendment. A second school is decidedly more critical of decentralisation in general and the 73rd Amendment in particular.

A central assertion that underlies this second school is that the reforms put in place in 1993 are largely cosmetic, and that the ultimate distribution of power and resources in rural areas is dependent on the pre-existing pattern of social inequalities created by caste, religion, class, gender and other forms of rural domination. This second body of thought as suggested, address in due course. For the time being however, we consider the extent to which the changes put in place by the 73rd Amendment can be understood as decentralisation.

Critical Assessments

Using the conventional classification of 'political, administrative and fiscal decentralisation,' the World Bank's three-volume study of Indian decentralisation ranks India 'among the best performers' internationally in terms of political decentralisation, but 'close to the last' in terms of administrative decentralisation.

Most States have held at least one round of elections since 1993. Reservations allowing the participation of women, Scheduled Castes and Scheduled Tribes have been respected.

Finally, voter participation has been high. In its study of 53 villages in Rajasthan and MP, for instance, the World Bank found that voter turn out in Panchayat elections was well over 90% for all categories.

This is significantly higher than the turnout for the most recent round of Lok Sabha elections, which was 61% for women and 65.9% for men.

The World Bank study goes on to argue that although Indian States and the Union government have been willing to recognise the Panchayats, to hold elections and to respect stipulations governing reservations for Scheduled Castes, Scheduled Tribes and women, they have been unwilling to vest them with sufficient 'administrative control over significant functions or fiscal autonomy'.

In most States, Panchayats have been handed a wide array of responsibilities without the necessary fiscal and administrative resources. In its assessment of Indian decentralisation, the Task Force on Devolution of Powers and Functions upon Panchayati Raj Institutions found that 'most of the States' had satisfied only the basic requirements relating to the transfer of functions, functionaries, funds and financial autonomy to the Panchayats.

Oommen's comparison of 12 Indian States concludes that the Conformity Acts have generally 'been an exercise in amending existing Panchayat legislation for the sake of satisfying the mandatory provisions of the 73rd Amendment.'

In particular, he raises the following concerns:

- Village Panchayats have been delegated functions 'without adequate administrative, financial and technical support,';
- With the exception of Kerala and West Bengal, Panchayats lack discretionary powers over spending and staff;
- There is insufficient clarity and differentiation of functions among Panchayats and other levels of government;
- States reserve the right to assign or withdraw functions to and from the Panchayats by 'executive fiat,';
- Panchayats at all levels have inadequate powers of taxation;
- Panchayats lack autonomous budgeting powers.

Similarly, Vyasulu finds that State governments have devolved little finances and fiscal powers to the Panchayats. Instead, many have established 'parallel bodies' as a channel for development funding. Self-help groups connected to the Janmabhoomi programme in Andhra Pradesh and the Rajiv Gandhi Watershed 'Missions' in Madhya Pradesh are two illustrations of this trend. Such findings are reflective of three more general problems, which are commonly associated with Indian decentralisation.

First, Indian States–and the government bodies that operate within their jurisdiction–are part of a federal system, in which powers are defined by a written Constitution, and divided among Union, State and sub-State bodies.

Second, the institutions empowered by Indian decentralisation are situated in a long-standing structure of public administration, whose interests and nature are not necessarily consistent with the provisions outlined in the 73rd Amendment.

Finally, this process has happened in a context of political transition, in which customs of discrimination and inequality in rural areas are thought to have been challenged by forces arising from the green revolution and other forms of guided intervention.

We consider the challenge of Indian decentralisation with respect to three 'institutional rigidities':

- The power of the States vis-à-vis the Panchayats;
- The nature and resistance of the Indian bureaucracy;
- Local élite 'capture'.

Federal Constraints

India's Constitution gives the States considerable powers governing the devolution of bureaucratic authority and the administration of agrarian institutions, such as land, land tenancy and agricultural labour.

The 73rd Amendment reaffirms this authority by giving the States great latitude in deciding the nature and scale of 'self-government'.

According to the 73rd Amendment, States are required to pass their own 'conformity legislation', which outlines the powers, functions and procedures of local government at village, district and intermediate levels.

However, the Conformity Acts also recognise the possibility that changing circumstances will give rise to new regulatory requirements, and therefore give individual States substantial autonomy to enact rules of 'delegated legislation,' which 'emerge through government orders and which in actual fact are the source of State control over the Panchayats,' Some idea of the extent to which the States of AP, MP, Kerala, Karnataka and West Bengal have retained powers of appointment, dismissal and review over the Panchayats: all States but Karnataka and West Bengal reserve the right to cancel decisions made by the Panchayats; all States but AP reserve the right to inspect the records of the Panchayats; all States reserve the right to dismiss the Sarpanch or village chief; finally, all States but Kerala reserve powers of appointment to the Panchayats.

Here it is worth emphasising that these are States that have been relatively progressive in the field of Panchayati Raj. A second requirement that emerges from the 73rd Amendment is the stipulation that the Governor of the State establish a State Finance Commission within a year of the commencement of the 73rd Amendment Act and every five years thereafter. The SFC is charged with the responsibility of reviewing the financial position of the Panchayats, and of recommending measures to augment the financial resources available to local bodies. Although many were quick to establish SFCs, few States have extended mandatory powers onto the SFC. A third important area of State authority relates to Panchayat elections.

The 73rd Amendment requires States to establish an independent State Election Commission whose primary responsibilities are to organise and oversee Panchayat elections, and to prepare the electoral rolls. Again, although most States have established SECs, many have been unwilling to relinquish powers of delimitation–*i.e.* the power to define electoral constituencies.

Beyond these three general powers, de Souza identifies five particular ways in which the States have retained control over the Panchayats:

- Through the application of law; *i.e.* Rules or executive decisions which undermine or over-ride the autonomy of local institutions;
- Through the removal, dismissal and assessment of local functionaries, whose ultimate accountability rests with the state government, not the panchayats;
- Through the cancellation of decisions or withholding of approval for panchayat activities;
- Through administrative means; *i.e.* The inspection of records, review of annual reports, etc.;
- Through the control of fiscal powers and resources.

As Jha has argued, one of the most important ambiguities relating to the functioning of the Panchayats is that of revenue generation:

- The decisions as to which taxes, duties, tolls and fees should be assigned to local bodies and which should be shared by the State with them are with the State legislatures. Consequently, while expenditure responsibilities of local bodies are likely to be extensively enhanced, there is no law to ensure a corresponding assignment of taxes or shares to match the additional responsibilities. Moreover, while the amendments did lead to the setting up of SFCs for improvement of fiscal scene at the local level, most SFCs chose to leave unchanged the existing tax powers of local bodies..

  Oommen classifies Panchayat finances into four categories:

  1. own resources;
  2. assigned revenues;
  3. grants; and
  4. loans.

  These can include a number of existing and potential sources of revenue:

  a. Taxes on land revenue, land, buildings, animals, and boats;
  b. Fees and user charges collected from the allocation of grazing rights, fishing rights, etc.;

c. Assignment of existing state taxes to local bodies;
d. Tax sharing with state governments;
e. Central government grants provided on the basis of the Central Finance Commission and the Planning Commission.

As of 2000, local governments in rural areas were not empowered to raise loans from public or private sources.

Jha assesses fiscal autonomy of the Panchayats in terms of:

- Their ability to obtain untied grants from the States and the centre and
- The requirement that they make a matching contribution to State or centrally sponsored schemes.

Reflecting on experiences in AP and Kerala, he concludes that many State grants are earmarked for a particular activity or constituency, and that autonomy of the Panchayats is minimal. AP, Karnataka and Rajasthan have all introduced the requirement that Panchayats make matching payments for particular schemes, such as irrigation works or streetlighting.

However, as international experience suggests, the ability to raise these resources is highly constrained by the size of the local revenue base and by the willingness and ability to impose taxes, levies, etc. on local voters. In most States, only village-level Gram Panchayats are empowered to levy taxes directly. As Jha points out, the 'higher level bodies are left out under the fear that tax powers to PSs and ZPs would result in tax base fragmentation and inter-jurisdictional conflicts'.

The question of fiscal autonomy is a matter of some debate. Some have argued that Panchayats do not necessarily need their own revenues to operate effectively. Access to 'guarded' or 'fenced' transfers from higher level authorities can serve this purpose, so long as the Panchayats have the autonomy to decide how the money gets spent.

Others, such as Oommen have argued that political autonomy is directly dependent on the ability to raise and spend public resources:

- Fiscal autonomy cannot be built into a regime of grants-in-aid. PRIs can play a genuine role in local development and self-rule only if they enjoy some degree of fiscal autonomy. This attribute is clearly more than relying on untied funds and rests on innovative resource mobilization, including project-tied loans, public contribution and the like. Tax assignment with clear taxing powers, and tax sharing assume significance in this context.

Leaving aside the problem of raising revenues in poor areas, there is also a problem of intergovernmental transfers. As Jha has argued, the State Finance Commissions have encountered difficulties assessing the financial needs of local bodies, reflecting both a lack of clear guidelines governing the function of SFCs and a dearth of reliable information on sub-district revenues and incomes. For the Panchayats, this has reinforced a situation in which local bodies have been handed vast responsibility with insufficient authority to raise and spend revenues and resources. In most States, the vast majority of Panchayat finances are provided either by New Delhi or by the State governments.

Central transfers from New Delhi to the States can take one of three forms:

1. Uied grants via the Finance Commission;
2. Pedominantly loan-based transfers from the Planning Commission/ Finance Ministry to the States for 'Plan' or developmental expenditures; and
3. Pedominantly grant-based transfers from the Planning Commission/ Finance Ministry to the Central Ministries for financing Centrally Sponsored Schemes and transfers to the DRDAs and the Panchayats.

In 2001–2, the combined transfer of funds from the Centre to the States was roughly 1,135 billion rupees. As a percentage of overall State expenditures, Central transfers dropped by around 18% between 1990 and 2001. This decline has been attributed in part to a decline in Central tax collection and to the deteriorating fiscal position of the Central government. The Finance Commission recommended that 100 billion rupees be allocated to the Panchayats and municipalities for the period 1999–2004. As Mukarji has argued, the prospects for self-government in India are greatly constrained by the political structures under which Panchayats are required to operate within India's federal system:

- More telling than any shortcoming of the Amendments is the all-pervasive mindset amongst the ruling élites that the natural mode of governing the country is through governments at the union and state levels only, as originally conceived by the founding fathers, and any attempt to insert a third stratum below the states would be fraught with unpredictable perils for the polity. The support for this view comes from the fact that the Constitution describes India as a Union of States, that its federal scheme divides powers between the union and the states, and that the union-states duality runs through its entire content... the 73rd and 74th Amendments make no consequential change in the pre-existing two-layered scheme of government outlined in the Constitution.

The tensions that strained India's federal system during the 1970s and 1980s have some bearing on the ways in which decentralisation has played out in the 1990s. Specifically, the rise of regional political parties and the more general feeling of resentment towards Mrs. Gandhi's strong arm tactics during the Emergency created a political context in which any weakening of the States would be perceived as an attempt to expand or consolidate the power of the Centre. Reinforcing this was a Constitution that conferred substantial power to the centre at the expense of the States. Although the relative powers of the states can of course vary with the resources they have at their disposal, the potential threat that decentralisation poses to the States cannot be underplayed.

As Mukarji has argued:

- The framers of the Constitution opted for a federation with a strong centre. Subsequently, the logic of the 'strong centre' concept shaped Congress politics and the policies of central planning in such a way as to make the union much stronger and, correspondingly, the states much weaker than the Constitution-makers ever visualised. If now the states endow meaningful powers downwards, the system will resemble an hour-glass, with the states at the narrow waist, the union power-wise broad at the top, and sub-state institutions function-wise broad at the bottom. Politically, this is unlikely to be acceptable to the states. Therefore, the union must first decentralize to the states so that they, in turn, may have enough to decentralize to lower formations.

In short, the political structures created during Independence and the political dynamics that unfolded in the 1970s and 1980s have combined to create a political environment that has been particularly hostile to Panchayati Raj. However, there are exceptions to this rule, and the exceptions can tell us something about the ways in which decentralisation can work in this relatively inhospitable climate. As in many fields of governance and development, the southern State of Kerala is a particularly interesting case in point.

In 1996, the Left Democratic Front government in Kerala launched the People's Campaign for Decentralised Planning, an exercise that resulted in the devolution of 35 to 40% of plan expenditure to local bodies. This involved unprecedented planning and coordination among the Gram Sabhas, as well as block and district-level Panchayats.

Although there were problems of co-ordination and some resistance from the non-elected bureaucracy, the process was notable both for the sheer scale of devolution as well as the political mobilisation that transpired around the issue of decentralised planning. Once again, however, it is vital to stress the fact that this process took place in a context of competitive party politics, in which the legitimacy of the ruling government was dependent on a re-distributive agenda. Moreover, this is a State where literacy levels for both men and women are in excess of 95%.

3

Sustainable Rural Development and Planning

RURAL DEVELOPMENT ACTIVITIES IN INDIA

The profile of livelihoods in India, especially in these states shows that agriculture and animal husbandry still remain the main sources of livelihood for rural communities, though there is a significant shift to non-farm sectors such as mining and quarrying, construction and manufacturing during periods of drought. The ultimate goal is for communities to have the confidence to make informed choices from a range of appropriate options that leads to sustainable and equitable development.

Socio-Economic Factors Affecting Rural Livelihoods

The poor have meagre holdings or access to land, little or no capital and off-farm employment is seasonal. It is almost impossible for farmers to secure credit and loans needed to purchase agricultural inputs except at prohibitive rates from private moneylenders leading to risk-prone farming. Markets are under developed or difficult to access. Extension services are few and far between, and development initiatives aimed specifically at their needs is sparse.

Few employment opportunities and low levels of education and skill result in low cash incomes. This in turn affects the ability to purchase basic needs (such as medicines, education for children, etc.). Women and children in particular are the hardest hit especially when access to safe and adequate sources of water are low, resulting in high vulnerability in terms of health. Women are also more affected by underemployment.

The relevance of this is all too clear when one realises that 29 percent of India's population still lives below the poverty line, earning less than US$ 1 per day. A comparision with other countries in South Asia reveals that 33 percent of Pakistan's population is below the poverty line while it 34 percent for Bangladesh, 42 percent for Nepal and 25 percent for Sri Lanka.

Major components of AKDN's rural development programme include: institution building, social organisation, natural resource management (especially in the area of water management and water-use efficiency including in coastal areas), productive infrastructure development and human resource development (especially of women), enterprise promotion, increased agricultural productivity and credit and savings services.

Promoting Self-Reliance

A central strategy of the rural livelihoods approach is to put people at the centre of development, thereby increasing the effectiveness of development assistance and therefore improving performance in poverty reduction. Involving the poor results in local empowerment, which in turn leads to opportunities for local leadership, including that of women, to emerge. These leaders play a critical role in bringing about and sustaining development and consequent social change such as pluralism, public participation and democratic principles. Village institutions form the basis of community organisation and function through 'Village Development Committees,' comprising of representatives of all communities in the village (including at least 30 percent women's representation). These Committees develop village development plans and work closely with the Panchayat (the lowest unit of the government that functions at the village level).

Other community organisations such as farmers' federations and women's self-help groups also facilitate the process of community driven development. These institutions work best if they have some independent means of sustaining themselves which is usually enabled by introducing them to micro-credit schemes. For example, a women's federation in Bharuch district in Gujarat state recently accessed a loan-based scheme for animal husbandry. Likewise, federations of farmers groups have considerably reduced agriculture input costs (seeds, fertilisers and pesticides) through bulk purchases based on demand from member institutions, simultaneously ensuring the quality of inputs supplied. These organisations have also contributed to improved cropping techniques, rational use of fertilisers and the adoption of appropriate low-cost technologies.

Communities eventually go on to form "apex institutions" (federations of various community institutions) at the block (a unit of a district) and "taluka" (a unit of a block) level. These institutions also act as forums where regional issues are discussed and solutions formulated. These institutions also dialogue with the government and other agencies (including banks) to get access to and benefit from various schemes.

The federations also serve as agriculture extension agents, and transfer information from agriculture institutes related to cropping practices, thereby ensuring that information reaches farmers in remote areas. Collective marketing of agricultural produce, and enabling member institutions to sell their produce at the best price possible is another important activity.

Farmers' federations and the women's federations have also led social campaigns in the regions such as reducing unnecessary expenditure on social customs, promoting the education of girls, anti-liquor campaigns, and the promotion of organic farming. Mass awareness campaigns by women's groups in Surendranagar in Gujarat for example, have led to a reduction in wasteful and extravagant expenditure on weddings and on social customs such as 'funeral feasts.'

Skills Development

In addition to institution building, programme activities also include human resource development including skills development to build up the skills base of villagers, especially women. For example, in the Netrang programme area of Bharuch district, women have been trained to repair and maintain hand-pumps. In another area women have been trained as masons. Communities receive organisation and financial management training to support the effectiveness and sustainability of village-level institutions, and key resource people are provided with technical skills to plan, implement and maintain the development activities. The ultimate goal is for people in the participating communities to have the access, confidence and competence to make informed choices from a range of appropriate development options.

In 2001 alone, over 300 training programmes benefited almost 6000 villagers, 64 % of which were women. These were essentially to increase awareness about savings and credit programmes, provide exposure to 'model' villages and orient rural communities to the principles of natural resources management. In addition, over 50 NGOs and 400 government staff have also been trained in various participatory processes.

Creating Rural Assets

Institutional structures are created at the village level through which the rural poor can prioritise their needs and decide how best to manage common resources. Communities build personal and community capital through efficient management of their natural resource base such as water storage and enhanced water use efficiency, irrigation systems, soil conservation or forestry.

These efforts include the construction of small scale infrastructure, such as check-dams, irrigation canals and water harvesting structures or agricultural storage facilities.

Over 400 structures have been created for harvesting and storage of rainwater that is directly lifted for irrigation, or recharging the ground water aquifers for more rational extraction through wells. These have led to an additional 4000 hectares of irrigated croplands in the programme areas, which is being further expanded through the adoption of water saving devices such as drip irrigation and sprinklers.

Income growth is promoted by increasing agricultural productivity through improved farming methods such as using drip-irrigation, provision of better seeds, creation and improvement of markets, land development, micro-credit, increasing off-farm incomes and supporting enterprise development. Local capital is mobilised by promoting savings and developing financial services to enable broad access to credit. Programmes are designed to have a combined effect so as to create a critical mass of economic activities that raise living standards.

In Surendranagar area in Gujarat for example, it was found that a severely degraded natural environment would not allow rural communities to rear cattle and therefore the only viable option was to rear goats. The rural development programme stepped in to help create a viable market for goat milk in the area and helped the community to get the government to set up a cheese making plant. In another water scare area, a women's federation lobbied with the government to buy them a water tanker which they now use to sell safe drinking water to local communities practically at their doorsteps and at fixed and mutually agree to prices. This has considerably reduced water scarcity in the area by providing households assured water supply and also saves women the time they earlier spent on collecting water.

Benefits and Impact on Quality of Life

Assets thus created often result in more than just increased income generation. Development activities have ensured more water for drinking and irrigation, higher agricultural productivity and rural incomes (including household savings), greater resilience to droughts and the ability to manage the natural resource base better.

With over 10,000 households now accessing assured sources of safe drinking water, women have benefited significantly. Since they no longer have to walk long distances to collect water, the time saved can now be spent with family or education of their children.

Alternatively they have more time for income generation activities. Health improvements due to improved diets and access to safe drinking water also lead to a reduction in the costs of medicines and trips to the doctor. Similarly, the number of days of illness has been reduced by 50 percent (down to 15 days per year, as compared to about 30 days per year earlier).

Simultaneously, literacy levels have risen by 10 percent (8 percent in the case of women). Growing gender sensitivity and changing gender equations is also evident from a phenomenal 350 percent increase in the expenditure incurred on medicines for women. School attendance levels show an improvement, especially in the case of girls who now no longer have to accompany their mothers at dawn to collect water.

Impact studies conducted in some areas reveal an average increase of beneficiary incomes by about 60 percent, going up to 80 percent in some cases. More savings and increased access to credit reduces the dependence of the poor on local moneylenders. Even the landless benefit from increased demand for farm labour when improved agricultural practices results in two or three crops instead of just one every year, leading to, among other things, a reduction in migration. In 147 villages in four states, savings amongst beneficiaries was estimated to be approximately US$ 0.6 million (28 million rupees approximately).

As rural employment has improved, stress-migration levels have shown a marked reduction (down to 35 percent from 80 percent earlier). Even the duration of migration has reduced to 90 days per year (down from 150 days a year earlier).

Sustainability and Partnerships

Long-term commitment is the key to creating sustainable impact. Consequently rural development programmes in India have now run 20 years and helped create replicable models that can be quickly adapted to a variety of contexts; as well as long-term relationships with donor agencies and peer organisations for the mobilisation of funds, human resources and expertise.

In terms of organisational elements, programmes are usually area-based programmes that serve a defined population, are implemented over a long time frame and at a significant scale and are characterised by comparatively thorough coverage of all of the population within their target area.

Ensuring sustainable impact requires consistent efforts over considerable time periods. In the case of rural development programmes, this has meant at least ten years of concerted efforts at the grassroots level before results and subsequent impact on the quality of life of the poor is visible for all to see.

AGRICULTURE IN INDIA

In FY 1987, field crops were planted on about 45 percent of the total land mass of India. Of this cultivated land, almost 37 million hectares were double-cropped, making the gross sown area equivalent to almost 173 million hectares. About 15 million hectares were permanent pastureland or were planted in various tree crops and groves.

Approximately 108 million hectares were either developed for nonagricultural uses, forested, or unsuited for agriculture because of topography. About 29.6 million hectares of the remaining land were classified as cultivable but fallow, and 15.6 million hectares were classified as cultivable wasteland. These 45 million hectares constitute all the land left for expanding the sown area; for various reasons, however, much of it is unsuited for immediate cropping. Expansion in crop production, therefore, has to come almost entirely from increasing yields on lands already in some kind of agricultural use. Due to lack of information about the land and their nature, we are still not getting the proper return with the yield of those lands.

The Community Development Programme in India was inaugurated in 1952 to implement a systematic, integrated approach to rural development. The nation was divided into development blocks, each consisting of about 100 villages having populations of 60,000 to 70,000 people. By 1962 the entire country was covered by more than 5,000 such blocks. The key person in the program was the village-level worker, who was responsible for transmitting to about ten villages not only farming technology, but also village uplift programs such as cooperation, adult literacy, health, and sanitation.

Although each block was staffed with extension workers, the villagers themselves were expected to provide the initiative and much of the needed financial and labor resources, which they were not in a position to do or inclined to do. Although progress had been made by the early 1960s, it was apparent that the program was spread too thin to bring about the hoped-for increase in agricultural production. Criticism of the program led to more specialized development projects, and some of the functions were taken up by local village bodies.

There was only a negligible allocation for community development in the sixth plan, however, and the program was phased out in the early 1980s. The Intensive Agricultural District Programme, launched in five districts in 1960 by the central government in cooperation with the United States-based Ford Foundation, used a distinctly different approach to boosting farm yields. The program operated under the premise that concentrating scarce inputs in the potentially most productive districts would increase farm-crop yield faster than would a wider but less concentrated distribution of resources in less productive districts. Among these inputs were technical staff, fertilizers, improved seeds, and credit. Under the technical guidance of American cooperative specialists, the program placed unusual emphasis on organizational structures and administrative arrangements. For the first time, modern technology was systematically introduced to Indian farmers. Within a decade, the program covered fifteen districts, 28,000 villages, and 1 million inhabitants. The Intensive Agricultural District Programme was thus a significant influence on the forthcoming Green Revolution.

IRRIGATION IN INDIA

Except in southeastern India, which receives most of its rain from the northeast monsoon in October and November, dryland cultivators place their hopes for a harvest on the southwest monsoon, which usually reaches India in early June and by mid-July has extended to the entire country. There are great variations in the average amount of rainfall received by the various regions—from too much for most crops in the eastern Himalayas to never enough in Rajasthan. Season-to-season variations in rainfall are also great. The consequence is bumper harvests in some seasons, crop-searing drought in others. Therefore, the importance of irrigation in India cannot be overemphasized.

Irrigation in India has been a high priority in economic development since 1951; more than 50 percent of all public expenditures on agriculture have been spent on irrigation alone. The land area under irrigation expanded from 22.6 million hectares in FY 1950 to 59 million hectares in FY 1990, an increase of 161 percent in four decades. This increase was about 33 percent of the estimated potential.

The overall strategy has been to concentrate public investments in surface systems, such as large dams, long canals, and other large-scale works requiring huge outlays of capital over a period of years, and in deep-well projects that also involve large capital outlays. Shallow-well schemes and small surface-water projects, mainly ponds (called tanks in India), have been supported by government credit but were otherwise installed and operated by private entrepreneurs. Roughly 42 percent of the net irrigated area in FY 1990 was from surface water sources. Tanks, step wells, and tube wells provided another 51 percent; the rest came from other sources.

Between 1951 and 1990, nearly 1,350 large-and medium-sized irrigation works were started, and about 850 were completed. The most ambitious of these projects was the Indira Gandhi Canal, with an anticipated completion date of close to 1999. When completed, the Indira Gandhi Canal will be the world's longest irrigation canal.

Beginning at the Hairke Barrage, a few kilometres below the confluence of the Sutlej and Beas rivers in western Punjab, it will run south-southwest for 650 kilometres, terminating deep in Rajasthan near Jaisalmer, close to the border with Pakistan. A dramatic change already had taken place in this hot and inhospitable wasteland by the late 1980s. As a result, desert dwellers switched from raising goats and sheep to raising wheat, and outsiders flocked in to purchase six-hectare plots for the equivalent of US$3,000.

Progress in irrigation has not been without problems. In India, arge dams and long canals are costly and also highly visible indicators of progress; the political pressure to launch such projects was frequently irresistible. But because funds and technical expertise were in short supply, many projects moved forward at a slow pace. The Indira Gandhi Canal project is a leading example. And the central government's transfer of huge amounts of water from Punjab to Haryana and Rajasthan, frequently cited as a source of grievance by Sikhs in Punjab, contributed to the civil unrest in Punjab during the 1980s and early 1990s.

Problems also have arisen as ground water supplies used for irrigation face depletion. Drawing water off from one area to irrigate another often leads to increased salinity in the supply area with resultant effects on crop production there. Some areas receiving water through irrigation are poorly managed or inadequately designed; the result often is too much water and water-logged fields incapable of production. To alleviate this problem, more emphasis is being placed on using irrigation water to spray fields rather than allowing it to flow through ditches. Furthermore, charges of corruption and mismanagement have been levied against government-operated facilities. Cases of bribery, maldistribution of water, and carelessness are frequently raised in the media.

Another major problem has been the displacement of thousands of people, usually poor people, by large hydroelectric projects. Critics also claim that the projects are damaging to the ecology. Smaller projects and such traditional methods for irrigation as tanks and wells are seen as having less serious impact. In the late 1980s and early 1990s, the debate between large-scale versus small-scale projects came to the fore because of the US$3 billion Sardar Sarovar project on the Narmada River.

Sardar Sarovar, as conceived, was one of the world's largest hydroelectric and irrigation projects. Some 37,000 hectares of land in Madhya Pradesh, Gujarat, and Maharashtra were slated to be submerged following the construction of some 3,000 dams, 75,000 kilometres of canals, and an electric power generating capacity of 1,450 megawatts of power per year. Included among the 3,000 dams was the proposed 160-meter-high Sardar Sarovar Dam.

In 1985 the World Bank agreed to loan US$450 million for the project. Environmentalists in India and abroad, however, argued that the project was ecologically undesirable.

In the face of this strong protest, the World Bank appointed a two-member team in 1991 to review the project. Despite a negative review of the environmental impact by the team, World Bank funding and the project continued. By 1993, however, in the face of continued international protest as well as opposition and a call for a satyagraha by villages in the affected areas, the central government cancelled the dam project loan. Work on the Sardar Sarovar project continues, however, with funds provided by the central government and the governments of the three states involved.

Although India had the second largest irrigated area in the world, the area under assured irrigation or with at least minimal drainage is inadequate. The irrigation potential estimated to have been created by the early 1990s was about 82.8 million hectares.

This amount includes the gross irrigated area plus the potential for double cropping provided by irrigation. There was a cumulative gap in irrigated land use of about 8.6 million hectares until FY 1990, by which time the gap had decreased through improved land management. Now days has come for the recycling of water and for this watershed shed management programme should be implemented properly.

RURAL LOCAL BODIES FOR FASTER AND SUSTAINABLE RURAL DEVELOPMENT IN INDIA

One of the first things that the East India Company did, by design, was the disruption of self-sufficient village economies, with interdependent social and agrarian relations, in India. The commercialization of agriculture and integration of village economy with the market, that served the interest of the British and middlemen, resulted in the breaking down of interdependence among the villagers and unemployment. The old benevolent privileged people of the villages became the collaborators of colonialism. The commercialization and breakdown of self-sufficient village economies impoverished the rural people and deteriorated their quality of life and livelihood. Their whole traditional system collapsed and got converted to the commercialized system; by integration with the market economy. The small landholders became the landless labourers.

Independence and After

The Panchayti Raj system was envisaged as reintroducing method of Gram Swaraj by empowering the people to deal with their own problems with the financial and technical support from the provincial and central government. It was the best possible effort at the time of inception. The ground realities in different states are mixed. There are states like Rajasthan, where Panchayats play major role in the democratic governance; and there are other states with varying degrees of empowerment.

Rural development and improving the quality of life in rural areas has been in force in our planning and developmental strategies all along. The numerous schemes that the Government of India has launched have had a sorry state of affairs in implementation.

The basic public services have not reached the rural areas especially the poor. The functioning of the systems is not encouraging the participation of all sections of the society. The transparency and accountability norms are hazy. The 'output' and 'outcome' of these schemes with waste expenditure have been poor.

The IRDP that was planned with good thinking had inherent weaknesses in terms of monitoring and evaluation. It is too common to be further stressed that the benefits monitoring was very poor and there were large scale leakages and inefficiencies in the schemes; that even today are a cause of concern. Nonetheless, it was the first major initiative in the right direction.

The recent programmes like the Bharat Nirman, PMGSY, SSA, SJGRY, BRDF, NFFWP, NREGP, PMGSY, SGSY; and NRHM have been launched with substantial funding. The effectiveness of these programmes will depend on the implementation by convergence and synergy.

Several of these programmes may have the overlapping and complimentary components; if that can be synergized, it will result in huge administrative cost cutting and efficiency in implementation. One can not wait in the infinity to village people acquiring the management expertise. It would rather be more fulfilling to make them learn by doing in the simplest, yet accountable, possible ways.

Development Unit for the Villages not Block

In most provinces, panchayats are viewed as agencies and not the institution for development and change management. Both, the politicians and bureaucrats, have shown sustained reluctance to make the panchayats as institutions of self government. Even in the 'spheres of activities' their local presence is not truly recognized in the Constitution of India; leaving the scope of interpretation open. To some extent, this is made up under the various State Acts; and Panchayats (Extension to the Scheduled Areas) Act, 1996. However, they have to depend on the states and the centre, for financial resources. The first thing that should be in place now, is the targeting the village panchayats as the nodal unit for the rural development and not the Block. The justification is simple-the development blocks have not proved effective agents of change in the villages!

It is also true that the awareness and the outreach of the basic infrastructure and communication technology are now in a position to communicate directly with the villages. This will ensure the direct access to the problems areas and remedial action both ways. Much of the failure of development schemes can be attributed to 'top-down' planning and approach towards the people.

District or Block level planning can be only generic; while the specific plan for the development of villages should be the domain of the Panchayat. District and block level establishment should be used for conceptual issues, facilitation and capacity building of Panchayats. They should have nothing to do with funds allocation, disbursement and projects. If they have the expertise they can do evaluation and impact assessment, independently.

Long Channels and the Hierarchy

First thing to facilitate convergence and synergy is to remove the long channels of implementation, disbursement and monitoring. Each channel adds to delay, inefficiency and other problems.

It is too well known a fact to be discussed here that in the present system there is always a gap between the funds required in terms of both quantum and timing. Much of this problem is due to the complex system of percolating down of the funds, possible leakages; and delays at each level. The remedy lies in empowering, educating and energizing the Panchayat System and allowing funds to be remitted direct to the Panchayats. The hierarchy within the Panchayat system (District, Block etc. levels) leads to 'external' and political influence; as well as other factors that undermine the role and responsibilities of the Panchayats! The simple systems get complicated, to suit everyone's vested interests in the hierarchy. The decision makers must be the local villagers and the Panchayats.

Community Empowerment

In the rural areas, the outreach of the administration is not as much as it is in the urban areas. Besides, the villages have common properties and assets in several ways. These are not owned by any single family or person. The community empowerment will ensure the effective participation and indulgence of the people in the spheres of activities of the villages and development.

One way to ensure the community empowerment is to allow social audit of the major activities and programs of the government and the panchayats. This will ensure the representative groups being the part and parcel of the planning, implementation of development schemes and programs, monitoring and evaluation thereof; and the assets management.

EMPOWERMENT OF PANCHAYATS

Education and continuing education is essential for any organization. It is heartening to note that most states have prescribed educational qualification for Gram Pradhans. These educated headmen have the requisite local experience and are well conversant with the basic problems of the villages. A *'Gram Pradhan'* is most accessible nodal point for every person in the village. In contrast with the MP's where there is no educational qualification required, the Gram Pradhans have had the benefit of education for furthering theirs views and ambitions in life.

The fact that, presently, Panchayats are more of a voluntary kind of institutions, it requires the attention of the government to make it as a career option for the villagers. Instead of spending millions on administering the schemes and programs at several levels, it would be prudent and more professional to make Panchayat jobs paid ones. The villagers and their representatives should have a major say in planning, project proposals and their implementation. In states where Revenue Police system is in place, they can be brought under the Panchayats for a better and effective system of dealing with law and order, revenue matters; and local disputes. With such empowerment, the Panchayats will require to be educated on management of priority schemes, financing, and monitoring oversight mechanisms. The capacity building of the Panchayats and the villagers in several ways can be organized by making the jobs look like simple ones and focused to their own village; or nearby areas in cases of common facilities and shared resources. Training and capacity building will energize the Panchayats and make them effective implementers of the plans, schemes and projects of which they have been a party. The whole system of Panchayat in the village will have to be energized by involving all sections of the population in the villages.

Low-Key Ombudsman

The monitoring of the effectiveness depends on good management information and accounts keeping. For purpose of accounts keeping regular position of an accountant can be created. The major concern that we may still have is to keep a proper vigil and watch on the speed and transparency.

e-Governance

Records keeping and retrieval of information for the land records, common properties, demographic information, assets etc. is generally poor in cases of all the villages. Sometimes, the poor and illiterate people do not even have access to such information. There is a growing need for use of information technology for possible solutions in the area. Wherever possible, e-Governance should be supported.

It is a myth that right now it is too early for the e-Governance in the rural areas. There are the problems of skills; but the same can be enabled by hiring people or associating the NGO's.

A case in example is the Ernakulam district panchayat in Kerala, who has taken up a unique model of e-governance to establish information networks as part of their plan projects; with a view to strengthen the decentralised planning system. However, due care has to be given to the fact that most rural population will require handholding and assistance in the new scenario. The stems in place will have to very responsive to the concerns and queries of the villagers.

Convergence of Schemes/Programs

The convergence of all ministry specific schemes in the Panchayat's Plan for implementation should have an overriding provision to reprioritize the schemes during the financial year by utilizing the available resources, for local modified needs. Here we are presuming the oversight and monitoring mechanisms enabled in the system.

The rural knowledge bank available with the senior citizen will come handy to decide as to what is more important for the village as a whole at any given time. In any case, the canalizing funds directly through the Panchayats will result in synergy and economy by containing the wasteful expenditure incurred in the name of administering the grants for the schemes. The resultant reduction in government jobs can be utilised by redeployment of skilled staff in the Panchayats; and as independent consultants operating privately. The system must ensure that there are *ex ante* targets fixed for *ex post* results and benefits monitoring, quantified in both physical and financial terms.

Local modifications in Schemes/Programs

Local modifications may be allowed to cater to the needs of the varying geographic conditions, allowing for the cultural factors; and better value for money derived. The housing or the water supply projects can never have the same specifications in the mountain areas of North-East or the middle Himalayan states. The development of local specific and user friendly technology in the projects will help achieve the goals better than any standard scheme or projects for all states.

Social Solutions and Holism

The dynamics of interrelationship of land holding, agricultural growth, employment; and poverty have to be understood in holistic way and strategies have to be devised to tackle them in tandem. People–centric and people oriented methods will entail the participation of the civil society, self-help groups, stakeholders, NGO's and other institutions to help achieve sustainable rural development with a clear vision and focus.

Fragmented and piecemeal efforts will only remain attempts without being translated into outcomes; and a burden on public exchequer.

Several of the complex problems do not require economic or political solutions; but the social solutions. We seem to be going overboard in bringing in more and more economic and political solutions, at the cost of neglecting the social solutions. The social changes are slower than even the slow process; but can be brought out only by tackling the social elements in association with the other measures and variables. The core areas of thinking shall be the holism of rural development:

"But holism requires much more – it must include the sense of all inclusive totality. Interestingly, this emphasis on all inclusive totality places at the disposal of less developed societies, an inexpensive volume of relevant R&D at the highest level of the corporate world which is struggling hard on the same problem of development – how to change the ongoing in order to survive in the world of future.

Why? Complex dynamic systems, whether physical, biological or societal, have the same underlying internal dynamics and exhibit common emerging patterns. All that is needed is the *cultivation and development of re-investment skills."*

Financial Management

Financial management and records keeping of any system is very important. We cannot expect the local bodies to perfect them without any regular support staff and system. Right from planning to outcomes reporting, skilled staffs are required. As said, elsewhere in this chapter the management of Panchayats should be a career option for the villagers; and the requisite internal expertise has to be available in the organization. Panchayats must have their own Budget and some sources of Income, by having powers of levy taxes, user charges etc. in the local sphere of activities. This will automatically involve everyone in the village administration, through the concerns of the tax payers and the users of services.

While the bulk of the sources of expending will continue to come from the State and Center's funds, the funds raising powers of the village panchayats will make them more aware of the areas that can be governed more effectively at the village levels. The maintenance of various existing and future assets that always starve for funds, will also find a new source for this purpose.

The large paraphernalia of the multi-tier system will provide the surplus manpower for redeploying in the Panchayats. It will in fact result into savings for the government in several ways. To begin with the panchayats may keep their cash based accounts on the 'single entry system', that reveals much more information on the transactions than the double entry or accrual system. The systems have to be made simple and transparent, unlike the complicated 8 Pallavi Dhyani (2003), Technological nurseries for rural industrialisation an d entrepreneurship data that is sought under the development schemes at present. In a fully accountable and transparent system, enabled with internal controls and audit, micro-management must be left to the Panchayats and only monitoring and evaluation should be the concern of the State and Central governments.

Internal Audit

An institutionalized mechanism of internal audit by independent agency/ NGO or even a kind of social audit system (of course making it simple) can be put in place. Most government institutions in the provincial and district administration also require having similar arrangement to make them accountable for their decisions.

Learning from Others

The empowerment of the rural local bodies will ultimately ensure that the 'ruralliving' is more prestigious than living in urban areas. In most European countries, living in the countryside is considered more prestigious than living in cities. Countries like U.K., Austria, Switzerland, and France are telling examples of that. Why they have such a situation is due to the comparable or better schools, hospitals, playgrounds, water supply, roads and general environment. In Alpine countries of Europe, there have been trends of reverse migration, with the upgraded rural livelihood.

Reinforcement of national programs and activities through decentralized decisionmaking and capacity-building has been undertaken by most of the countries in the Eastern Europe. Activities range from development of farmer support groups to the establishment of central and regional development councils, as in Hungary, which are responsible for regional planning, priority setting and allocation of financial resources for various regional development projects. Through the Estonian Village Movement, rural communities have been active in appealing to the Estonian Government for assistance in rural development issues. The first Rural Assembly for Villages was held in Estonia in 1996 and concluded with the formation of the Estonian Movement of Villages and Small Towns (KODUKANT), a nongovernmental organization.

The international community recognizes that concerns about the impacts of globalisation and the free market, nutrition, supply of safe drinking water and energy, and improvements in education and health. Government measures would go a long way to solving problems of poverty, but improved coordination, cooperation, and targeting of investments was needed. The unique characteristics could be exploited to capture niche markets for products and services such as organic food products and eco-tourism.

The availability of such amenities and infrastructure enables the rural tourism to compete strongly with the city tourism. One the rural tourism catches up, the rural people will find jobs in their backyard of their ancestral village home; the pain of cutting the umbilical cord will also be minimized. This sounds very simplistic but requires a lot of investment, will and long term planning. Cities as 'Engines of Growth' will provide jobs and income; but certainly not rural development as most investment analysts think.

TECHNOLOGY AND MECHANIZATION INDIA

Despite the pervasive, large-scale use of draft animals throughout India, agricultural machinery and implements, tractors, in particular, have had an important place in increasing agricultural productivity. The stock of tractors increased from 8,600 in FY 1950 to 518,500 in FY 1982 and continued to grow rapidly throughout the 1980s. The number and sale of power tillers and combine harvesters produced and sold was small, with 4,678 tillers and 110 harvesters sold in FY 1988. There was a significant increase in the number of electric pumps and oil pump sets for irrigation during the 1980s. The production and use of machinery are hampered by the small size of many operational holdings.

However, a number of improved agricultural implements are available for tilling, seeding and fertilizer application, weeding, harvesting, and threshing. The implements include moldboard plows, disc harrows, cultivators, seed drills (more than 110,000 were sold annually in the early 1990s), and mechanical power threshers. These tools have the potential of increasing yields for all crops, but the adoption rate of improved machinery is low.

The Central Institute of Agricultural Engineering at Bhopal, Madhya Pradesh, under the aegis of the Indian Council for Agricultural Research, is responsible for coordinating the manufacture and promotion of technology for small and marginal farmers. The government introduced an incentive scheme in 1990 to subsidize the cost of machinery by up to 50 percent to small and marginal farms. Additionally, farmers' agro service centres are being established to provide custom service for improved implements and machinery. The eighth plan includes a major thrust for upgrading and adopting proven technology.

In a country with a large and growing labor force, too much mechanization in the short run could create fossil fuel shortages as well as social and economic problems. There is, nevertheless, room for improvements in technology. Since FY 1983, there have been attempts to popularize improved animal-drawn agricultural implements and hand tools through demonstrations and subsidies to small and marginal farmers.

Despite these advances in mechanization, most crops in India are still sown, transplanted, weeded, and harvested through labor-intensive human work. Most grains in India are harvested by teams of Indian labourers wielding hand-forged iron sickles, binding up sheaves of grain, and carrying loads of sheaves on their heads to bullock carts to be transported to threshing floors. Teams of bullocks are then driven over the sheaves to separate the grains from the stalks, and workers toss basket loads of grain into the air to separate the grain from the chaff. Lentils, a crucial part of the Indian diet, also are harvested through labor-intensive means. Groups of labourers squat down in fields for hours at a time, ripping out lentil plants at the root by hand. Machinery available to lentil farmers has proven difficult to use, and traditional methods are preferred.

Land Utilisation

Major changes in the land use systems of any District, there is an urgent demand for adaptive environmental assessment. Semi-arid climatic conditions together with high land use pressure from intense small-scale settlements cause frequent crop failures, overgrazing and erosion damage. A spatial information system can be developed to handle the unique database demands for this study, and for further District planning purposes. Soil mapping, using GIS as a modelling tool, helped to improve existing and augment incomplete, poor or even absent data, such as a soil map on a reconnaissance level. Despite a semi-quantitative approach, modelling of the multivariate relations of risks reflects their spatial distribution clearly. Land consolidation is the most efficient possibility for land property policy and the most generally-used application. It is an administrative procedure used to reduce the number of property plots, in order to improve the relation between ownership and land use, in the interest of economic production. The topic which concerns the relationship between land ownership and land use, was handled with special care in different economic and political systems. The Indian Land Property policy varies state to state, which control the property market, land reforms and land allotment. Land property policy of India should be made unique and then only it will have efficiency also in the environmental and nature protection, in the extensive agriculture and the formation of environmentally-sensitive zones as well. First and last the aim of the new land consolidation concept is to define the relation between land property and land use. This consolidation makes it possible to form reasonable sized agricultural farms with the help of private farming and lease-firm and also it improves the ownership structure using national funds for supporting.

GIS BASED APPLICATION FOR RURAL DEVELOPMENT

Geographic Information Technology has developed at a remarkable pace over the past two decades and will play a key role in development of nations in the 21st Century, thereupon many countries have already prepared their strategic development plans for application of GIS Technology with gigantic financing endeavors. Now time has come for all decision makers to discuss the appropriateness of GIS technology and its applications to rural development, forest management, urban development planning, land information systems and agricultural development. This will also provide a suitable solution for the use of GIS for educational infrastructure development with special emphasis on rural sector in India.

Educationists, planners, researchers, decision makers, administrators, communication professionals and officials from different departments and some reputed NGOs should be invited to discuss the role of GIS Technology and should implement the same outcome immediately for ensuring sustainable development and socioeconomic and educational uplifting of the country.

Information Technology has emerged as an inevitable phenomenon influencing every walk of life of people in all sections of this society. With the ease of availability of enormous computing power and convenient access to large volume and variety of data and information, the structure and functions of all human organisations will undergo profound transformation in this century. Nations are engaged in exploiting this phenomenon for many of their socioeconomic requirements. One area, which is engaging serious attention, relates to use of Information Technology in the governance systems, especially the tools and techniques for acquisition and management of data relating to Geographic Information System (GIS).

GIS IN INDIA

India maintains a pre-eminent position in the use of spatial imagery. The capabilities in the development of high-resolution satellites and extensive network of associated infrastructure have contributed to the growing interest in the application of GIS for a variety of India's development needs. Indications are that these applications will continue to grow even more rapidly in the coming years. Since the spatial imagery is becoming easier to use and more affordable, the user base for GIS is expanding in several directions in seeking holistic solutions beyond image processing capabilities. The Information Technology policy of Government of India adopted in 1999 emphasizes the availability of spatial data to GIS user community and industry, thereby enabling the widespread development of Spatial Decision-Support Information System Network including Web enabled GIS application services. The Indian export from GIS segment is expected to increase to US $ 150 million in the next five years from the present level of US $ 60 million. The areas which are receiving priority attention include natural resources information assessment, monitoring and management, water shed development, environmental planning, urban services and land use planning.

Most States in India and several ministries and departments of the Central and State Governments have initiated special GIS programmes relating to ground water studies, cadastral mapping, power transmission and transportation infrastructure. The integration of socioeconomic data with spatial data is increasing.

The institutional infrastructures have been developed across the country catering to the local, regional and national needs. Some of the institutions with sophisticated capabilities are: National Remote Sensing Agency, Hyderabad, Indian Institute of Remote Sensing, Dehradun, Space Application Centre, Ahmedabad, Regional Remote Sensing Service Centres at Bangalore, Nagpur, Jodhpur, Institute of Remote Sensing at Anna University, Chennai and Survey of India Training Institute, Hyderabad. These institutions offer a variety of training programmes relating to GIS besides undertaking or supporting large scale application projects.

At present a large number of private firms of Indian and foreign origin have been active a undertaking GIS projects. They have been particularly responsible in introducing in the country instruments, softwares and educational programmes. Central Mining Research Institute (CMRI), one of the premier laboratory of Council of Scientific and Industrial Research (CSIR) is fully equipped with latest IT related infrastructure and their Scientist are capable for undertaking any GIS related projects of India and abroad.

NIC is also providing assistance to several Central, State and Local bodies in fulfilling their specific GIS requirements. In recent years most of the organisations engaged in GIS activities have felt the need for establishing control points especially in applications such as land records management, cadastral survey and hydrographic survey.

For these tasks, acquisition of Global Positioning System (GPS) has been increasing. During the year July 2000-June 2001 about 50 GPS equipments have been procured by several governmental agencies in the country. Despite these noteworthy achievements in GIS in India, there are still some limiting factors that need to be addressed, such as the restrictions on the availability of high-resolution data in sensitive areas, lack of nationwide control points, absence of more convenient repository and retrieval systems and lack of standardization of map scales.

The implementation of GIS in Research Programme raised a variety of conceptual questions for both the ecological and the socioeconomic sectors of this regional, integrated research programme. In addition to these basic units of research, spatial links between the two sectors and levels of data abstraction for the spatial database had to be defined. Using the theoretical background of the hierarchical system approach and valuable experiences of spatial data handling a consistent spatial information database can be created. Despite problems with data accuracy, logical consistency and completeness of data, a powerful tool for regional and local planning can be developed which can serve as a framework for a variety of planning purposes at the local and regional levels, as well as the transfer of know-how between governmental agencies and institutions using an interactive approach.

ROLES OF NGO ACCORDING TO THE EXPECTATION OF PEOPLE

NGOs play a critical role in all areas of development. People and policy makers are agree on one thing that NGOs play a very important role in development. Role of NGOs vary over the years as the policy of government changes. NGOs are almost dependent on polices of government. Socio economic development is a shared responsibility of both i.e. government and NGOs. Role of NGOs are complementary but vary according to polices of government.

If we closely pursue the voluminous literature on NGOs many roles can be found according to the expectations of people. The major development roles ascribed to NGOs are to act as:

- Planner and implementer of development programmers,
- Mobiliser of local resources and initiative,
- Catalyst, enabler and innovator,
- Builder of self reliant sustainable society,
- Mediator of people and government,
- Supporter and partner of government programme in activating delivery system implementing rural development programmes, etc.,
- Agents of information,
- Factor of improvement of the poor, and
- Facilitator of development education, training, professionalisation, etc.

Basically NGOs role is to prepare people for change. They empower the people to overcome psychological problem and opposition of oppress. Its role cannot be denied.

OBJECTIVES OF NGOS IN DEVELOPMENT

NGO is one of the alternatives available among various development organizations and one of the inputs among technical, financial and other resources, major merits of NGOs are emerging from their limited scale of operation; the sporadic efforts of NGOs can be consolidated and made more effective. Still the primary role of NGO is at the local level as mobilizes of people and their resources for an indigenous self-sustainable development. And at this level it can be a pioneer, mediator power broker, catalyst and has many other roles. NGOs and their long lomerations also are very in playing their role as advocates in policy issues beyond local level-national or even international level. Proper assessment of expected an actual roles of NGOs enable us to make them an effective alternative in the development process. However, small and sporadic NGOs are, they are valued in a pluralist society as an alternative approach to conventional system of attaining human well being and as such NGOs have a pivotal role to play in any society especially where institutions are alienated and development is dehumanized.

An NGO is nowadays not expected to deliver directly some benefits to people, but to motivate people, mobilize resources, initiate leadership, and participate in development programmes for self reliance. An NGO is only an enabler and as and when a society is made self reliant, role of NGO is shifted to another place where NGO service is required. But NGO works in relief and delivery of public goods as direct suppliers and majority of the development NGOs are also involved directly in productive activities. The roles of NGOs an enabler or catalyst for self reliant society and as supplier or implementer is relevant where bureaucracy is indifferent or inefficient, programs lack flexibility and cost effectiveness poor are ignorant, elite are ambitious, successes and services are pre conditions for motivation etc.

TYPES OF NGOS

Types of NGOs can be understand by their level of orientation and level of cooperation. Types of NGOs by the level of orientation. It has further types as under,

i. Charitable orientation. It often involves a paternalistic effort with little participation by "beneficiaries". It includes the NGOs which directed the people towards meeting the needs of poor and help them by gaining them food, clothing, medicine, provision of housing etc. such NGOs may also undertake relief activities during natural or man made herds.

ii. Service orientation. It includes with NGOs with activities such as the provision of health, family planning or education services in which the program is designed by the NGOs and people are expected to participate in its implementation and in receiving the services.

iii. Participatory orientation. It is characterized by self-help projects where local people are involved particularly for example in the implementation of a project in any village by contributing, cash, tools, land, materials and labor etc. this type is basically cooperation based and on limited scale.

iv. Empowering orientation. The aim of these NGOs are to help poor people an d develop a clear understanding of the social, political and economic factors which are effecting their lives, and aware them how can they solve their problem by using their resources and purpose to mobilize the people or self mobilization. In any case there is maximum involvement of the people with NGOs acting as a facilitators.

Types of NGOs by the level of operation. It has further types which are as following.

i. Community based organization (CBOs) When people start feelings that what are their needs and how can they fulfil them. These NGOs arise out of people's own initiatives. These can includes sports clubs women organizations neighbourhood organizations, religious and educational organizations. Some supported by NGOs, national and international NGOs and other independent outside help. Some are devoted to raising the consciousness of urban poor or helping them to understand their rights in gaining access to needed services while others are involved in providing such services.

ii. Citywide organizations. These NGOs are organized for some major or personal purpose. For example cambers of commerce and industry, coaliation of business, educational group. Some exist for other purposes and become involved in helping the poor as one of many activities, while others are created for the specific purpose of helping the poor.

iii. National NGOs. It includes organizations such as the Red cross, YMWCAs, YWCAs, professional organizations etc. Some of these have state branches and assist local NGOs.

iv. International NGOs. These range from secular agencies such as REDDA BARNA and save the children organisation, CARE, UNDP, UNICEF. Their activities vary from mainly funding local NGOs institutions and projects and implementing the projects themselves.

NGOS EMERGENCE IN PAKISTAN

In Sub-continent NGOs culture took shape in the form of Ghandi Ashram Banaras in 1927. This venture created to provide jobs for natives in the days of British Raj Diyal Sigh trust is an example of the time but these efforts could not become a social norm due to a highly centralized bureaucratic governance. As it is not a new phenomenon for Sub-continent. It emerged during the colonial period when religious, linguistic and ethic communities felt their cultural, religious and social identity threatened. Renouncing politics they concentrated on religious, cultural and social assertion.

To control these associations, the colonial authorities introduced the system of registration under the act of cooperative societies each society was required to give constitution and by laws and maintain financial accounts. The major purpose of theses organizations was to open educational, institutions, help the poor and destitute and improve the condition of women. Such welfare, charitable and educational organizations produced a breed of social workers who devote their lives to social work. They were sincere and concerned with the welfare of their community. These community based organizations also created a since of competition among each other which resulted in positive development. The great contribution of old NGOs was that they preserved cultural, social and religious values and in resistance to colonial states started movements which lead to positive struggle in the field of development.

Unfortunately, in Pakistan these NGOs cannot work properly due to political weakness' till 1958 so, under these circumstances in 70s new type of NGOs emerge which were quite different from old ones. The only thing common in old ones and new NGOs is that both came into being into response of state weaknesses. But in the absence of well-defined policy for NGOs, there is no moral considerations practiced by the people who have monopolized this vital sector. On one hand they serve as an employment exchange for kith and kin of the privileged and on other hand they are a symbol of prestige for the selected few that know the art of preparing proposals and report written in the bureaucratic lexicon. NGOs claim that there only task is to create social awareness but when people fell that NGOs are not helping them concretely, they lose interest and merely social awareness is of no use to them. Some of the clever participants turn this opportunity to their own favours by manipulating different NGOs to get funds in the name of social work. They know that projects are foreign funded and there is no commitment and sincerity behind it. Most NGOs have more or less become family business making big profit if you are a good pretender you can generate huge funds. As it is discussed above that NGOs receive funds from broad but nobody knows where and how these funds are utilized. So, people don't trust NGOs foe help as they consider them as fraud.

NGOS AND THEIR ROLE IN THE GLOBAL SOUTH

Turner and Hulme (1997) define NGOs as "associations formed from within civil society bringing together individuals who share some common purpose." Hulme (2001) characterizes them (as well as civil society) as "peopled organizations [that] are both not part of the state structures, are not primarily motivated by commercial considerations or profit maximization, are largely self-governing, and rely on voluntary contributions (of finance, labour or materials) to a significant degree." So, as Fowler (2000) observes, "for our purpose, business is not included." In support of this analysis, Edwards and Hulme (1995) expound that "most organizations referred to as NGOs thus belong, analytically, to the private sector, albeit to the service (i.e., not-for-profit) sub-sector thereof."

From a larger perspective, the prominent elements of a society are the state, the market (private sector), and the civil society – the "third sector." The state maintains public order and, to one degree or another, serves its citizens' needs. Companies in the marketplace pursue profits. And NGOs? They, like the state, seek to serve community needs, such as health, education, water, and sanitation. They may do so at least partly with government funding. For example, Clark (1991) states that "Indian NGOs are now of much significance to the country's development efforts.... The government's latest five-year plan has a one and a half billion rupee provision for funding NGOs." At the same time, though, NGOs stand apart from the state, engage in policy advocacy, and sometimes criticize government institutions and officials. As Fowler observes (1997), NGOs may provide a link between micro-level actions (providing individuals or communities with construction materials, farming equipment, or legal advice, for example) and macro-level actions (policy advocacy, lobbying, and monitoring how the state uses its powers).

Micro and macro can be viewed from a different perspective, as Turner and Hulme (1997) do: "a primary distinction can be made between organizations that are based in one country (or several countries) and [those that] seek to assist in the development of other countries. These are international NGOs (INGOs). Intermediate NGOs which operate across developing country or a region of a country [in the South can be termed] Southern NGOs. Closest to the practice of development are grassroots organizations (GROs) that operate within a limited area such as in a group of villages or part of a city."

DEVELOPMENT PLANNING

Planning in India dates back to the 1930s. Even before independence, the colonial government had established a planning board that lasted from 1944 to 1946. Private industrialists and economists published three development plans in 1944. India's leaders adopted the principle of formal economic planning soon after independence as an effective way to intervene in the economy to foster growth and social justice.

The Planning Commission was established in 1950. Responsible only to the prime minister, the commission is independent of the cabinet. The prime minister is chairperson of the commission, and the minister of state with independent charge for planning and program implementation serves as deputy chairperson. A staff drafts national plans under the guidance of the commission; draft plans are presented for approval to the National Development Council, which consists of the Planning Commission and the chief ministers of the states. The council can make changes in the draft plan. After council approval, the draft is presented to the cabinet and subsequently to Parliament, whose approval makes the plan an operating document for central and state governments.

The First Five-Year Plan (FY 1951-55) attempted to stimulate balanced economic development while correcting imbalances caused by World War II and partition. Agriculture, including projects that combined irrigation and power generation, received priority. By contrast, the Second Five-Year Plan

(FY 1956-60) emphasized industrialization, particularly basic, heavy industries in the public sector, and improvement of the economic infrastructure. The plan also stressed social goals, such as more equal distribution of income and extension of the benefits of economic development to the large number of disadvantaged people. The Third Five-Year Plan (FY 1961-65) aimed at a substantial rise in national and per capita income while expanding the industrial base and rectifying the neglect of agriculture in the previous plan. The third plan called for national income to grow at a rate of more than 5 percent a year; self-sufficiency in food grains was anticipated in the mid-1960s.

Economic difficulties disrupted the planning process in the mid-1960s. In 1962, when a brief war was fought with China on the Himalayan frontier, agricultural output was stagnating, industrial production was considerably below expectations, and the economy was growing at about half of the planned rate. Defense expenditures increased sharply, and the increased foreign aid needed to maintain development expenditures eventually provided 28 percent of public development spending. Midway through the third plan, it was clear that its goals could not be achieved.

Food prices rose in 1963, causing rioting and looting of grain warehouses in 1964. War with Pakistan in 1965 sharply reduced the foreign aid available. Successive severe droughts in 1965 and 1966 further disrupted the economy and planning. Three annual plans guided development between FY 1966 and FY 1968 while plan policies and strategies were reevaluated. Immediate attention centered on increasing agricultural growth, stimulating exports, and searching for efficient uses of industrial assets. Agriculture was to be expanded, largely through the supply of inputs to take advantage of new high-yield seeds becoming available for food grains. The rupee was substantially devalued in 1966, and export incentives were adjusted to promote exports. Controls affecting industry were simplified, and greater reliance was placed on the price mechanism to achieve industrial efficiency.

The Fourth Five-Year Plan (FY 1969-73) called for a 24 percent increase over the third plan in real terms of public development expenditures. The public sector accounted for 60 percent of plan expenditures, and foreign aid contributed 13 percent of plan financing. Agriculture, including irrigation, received 23 percent of public outlays; the rest was mostly spent on electric power, industry, and transportation. Although the plan projected national income growth at 5.7 percent a year, the realized rate was only 3.3 percent.

The Fifth Five-Year Plan (FY 1974-78) was drafted in late 1973 when crude oil prices were rising rapidly; the rising prices quickly forced a series of revisions.

The plan was subsequently approved in late 1976 but was terminated at the end of FY 1977 because a new government wanted different priorities and programs. The fifth plan was in effect only one year, although it provided some guidance to investments throughout the five-year period. The economy operated under annual plans in FY 1978 and FY 1979.

The Sixth Five-Year Plan (FY 1980-84) was intended to be flexible and was based on the principle of annual "rolling" plans. It called for development expenditures of nearly Rs1.9 trillion (in FY 1979 prices), of which 90 percent would be financed from domestic sources, 57 percent of which would come from the public sector. Public-sector development spending would be concentrated in energy (29 percent); agriculture and irrigation (24 percent); industry including mining (16 percent); transportation (16 percent); and social services (14 percent). In practice, slightly more was spent on social services at the expense of transportation and energy. The plan called for GDP growth to increase by 5.1 percent a year, a target that was surpassed by 0.3 percent. A major objective of the plan was to increase employment, especially in rural areas, in order to reduce the level of poverty. Poor people were given cows, bullock carts, and handlooms; however, subsequent studies indicated that the income of only about 10 percent of the poor rose above the poverty level.

The Seventh Five-Year Plan (FY 1985-89) envisioned a greater emphasis on the allocation of resources to energy and social spending at the expense of industry and agriculture. In practice, the main increase was in transportation and communications, which took up 17 percent of public-sector expenditure during this period. Total spending was targeted at nearly Rs3.9 trillion, of which 94 percent would be financed from domestic resources, including 48 percent from the public sector. The planners assumed that public savings would increase and help finance government spending. In practice that increase did not occur; instead, the government relied on foreign borrowing for a greater share of resources than expected.

The schedule for the Eighth Five-Year Plan (FY 1992-96) was affected by changes of government and by growing uncertainty over what role planning could usefully perform in a more liberal economy. Two annual plans were in effect in FY 1990 and FY 1991. The eighth plan was finally launched in April 1992 and emphasized market-based policy reform rather than quantitative targets. Total spending was planned at Rs 8.7 trillion, of which 94 percent would be financed from domestic resources, 45 percent of which would come from the public sector. The eighth plan included three general goals. First, it sought to cut back the public sector by selling off failing and inessential industries while encouraging private investment in such sectors as power, steel, and transport. Second, it proposed that agriculture and rural development have priority. Third, it sought to renew the assault on illiteracy and improve other aspects of social infrastructure, such as the provision of fresh drinking water. Government documents issued in 1992 indicated that GDP growth was expected to increase from around 5 percent a year during the seventh plan to 5.6 percent a year during the eighth plan. However, in 1994 economists expected annual growth to be around 4 percent during the period of the eighth plan.

Four decades of planning show that India's economy, a mix of public and private enterprise, is too large and diverse to be wholly predictable or responsive to directions of the planning authorities. Actual results usually differ in important respects from plan targets. Major shortcomings include insufficient improvement in income distribution and alleviation of poverty, delayed completions and cost overruns on many public-sector projects, and far too small a return on many public-sector investments. Even though the plans have turned out to be less effective than expected, they help guide investment priorities, policy recommendations, and financial mobilization.

PERFORMANCE OF INDIA"S PLANNED ECONOMIC DEVELOPMENT

We started our planned economic development since 1951 having in our hands the experience, a wide literature of well proved strategies and variegated plans pertaining to, used by and created or formulated by the well developed western economies, apart from our available natural resources. We were over enthusiastic and over ambitious on account of having the ready and well proved weapons sought from the western world for combating the problem of development. Therefore, instead of starting from the very beginning and covering the whole path we, being enticed and allured by the surprisingly fascinating fruits of industrialization, started our efforts but having longed for being developed and grabbing fruits thereof in a haste. Thus we lost sequences in our development path. We ignored agriculture that was the spine of our economy. Thereby our agricultural development lagged far behind the level required for feeding our industrialization up to the mark. Agriculture based small and cottage industries became shattered and the villages became ruined. This raised a huge bulk of unemployed people in the widely spread rural sector.

The unemployed persons started migrating to the urban areas in search of job. The urban development and industrialization there had not sufficient level to absorb the whole migrating mass and to provide them proper urban life facilities. As a result thereof a mushroom growth of slums came about fast which eventually turned into big slum spots in cities and towns within a no longer period of 20 or 25 years. This hampered urban growth and urban life. On the other hand, in rural areas there emerged acute shortage of energetic workforce, service centres, infrastructure, intellectuals etc. This hampered the rural development whereby agricultural development and rural life remained slang lower.

That is why, even having travelled a long path of planned economic development, the state of affairs of rural India still remains almost the same as it was before planning. There is a big gulf between urban and rural people regarding wealth, wage, education and income. Moreover, the rural-urban migration, due to the pitiable state of affairs in rural areas, resulted to unchecked urban growth. Thus, instead of overall development, an unbalanced and unfair development of our economy became resulted therein. Consequently a wide spread general unemployment prevailed in both the rural and the urban areas (as per ECONOMIC APPRAISAL 2006-07, the estimated number of unemployed persons rose from 7.98 million in 1983 to 9.02 million in 1993-94, to 10.51 million in 1999-2000 and to 13.10 million in 2004-05.

In addition to this a considerable number of politically, socially and economically sound and effective elites emerged in cities and urban towns. These elites interfered in the formulation and execution of development plans, on one hand, and in the fixation of priorities, on the other. Thereby our development plans became urban oriented and concentrating on rich minority.

Thus the poor majority and the rural economy became ignored. The unemployment situation in both the rural and the urban sectors became almost uncontrollable. The government became politically weak.

Therefore its priority became to please the rich minority so that it may run. To mitigate unemployment and poverty among the general mass it has to play pseudo role to remedy some times the rural and some times the urban mass alternately through various unsuccessful employment programmes and plans. Due to the general unemployment the condition of rural mass became more embarrassing than that of urban mass since the rural people had already been subsisting in privation. The fruits of the development programmes in rural sector were grabbed by the social elites in villages.

Therefore, the general rural mass, instead of observing their economic uplift, rather found them-selves lagging behind in the run of economic development. The village industries had to be liquidated; rural artisans or handicraftsmen either became unemployed or had to migrate to urban cities or towns; the socio-cultural, cheap and some times free entertainment sources (like village dramas, swangs, street magicians, community chorus, community games, rural fairs, community festivals etc.) were snatched by costly media based means. Modern road and rail transport replaced the traditional transport. Rural weekly markets and village shopkeepers came to their end in the want of adequate demand. This resulted to a rapid increase in the cost of living of the rural people but their income remained far behind.

CAUSES OF THE REGRESSIVE EFFECTS ON THE RURAL LIFE

There have been three causes of this stinging situation. The first was our galloping along the development path instead of travelling along the true locus by creeping walking and running as and when needed. The second cause refers to the distortion in the concept of rural development which made the formulation of development plans and strategies incompatible to rural economy. The third cause was the rapid population growth which added a lot to make the unemployment situation a mammoth. The first cause relates to our over enthusiasm and the second cause relates to the hereinabove discussed third theme of rural development literature. The third cause relates to the increasing difference of birth rate over death rate and the insignificant performance of'Family Planning Programme'.

On account of extended medical facilities, uplift of living standard due to increased national income, control over epidemics, check on famines, alleviation of starvation, extension of maternity services etc. during the development process in the plan period, the death rate considerably went down (from 27.4 per thousand per year during 1941-50 to 7.6 per thousand per year in 2005) but the birth rate remained slang high (it was 39.9 per thousand per year during 1941-50 and came down only to 23.8 per thousand per year in 2005). Therefore population growth attained an increasingly high rate that was however tried unsuccessfully to be lowered through the'Family Planning Programme'.

NGOS, DEMOCRACY AND SUSTAINABLE DEVELOPMENT

NGOs have rapidly gained an international reputation. They are being consulted on matters of policy making, planning and implementation at the local, national and international levels. This is because they offer new perspectives as well as a wealth of experience in such areas as grassroots development, environmental protection and the defence of human rights. They have a broad knowledge base and strong commitment to issues relevant to the emancipation of the poor. They have shown more sensitivity and understanding of Africa's severe economic and political crisis than their respective governments and the multilateral institutions. Keeping in mind that Africans at the grassroots level must have the lead in defining their needs and formulating development strategies, NGOs have recommended an action-orientated 'compact' for African development which must be translated into coordinated programmes for long-term solutions. Indeed, many NGOs are more active and knowledgeable in the area of development needs of poor countries than are their governments. They have demonstrated an awareness of the fact that conventional development models have not changed the situation of the poorest in Africa and they have been quick to recognize that people are poor because they have no power.

NGOs are increasingly influencing their governments, mobilizing people at the grassroots level by strengthening their institutions and raising their awareness; bringing to the attention of donor countries and the multilateral institutions the harmful effects of some of their policies; calling for equity-led strategies that give priority to achieving broad participatory ownership, control and management of natural resources by people to serve their own needs; and decrying current development strategies which favour the market economy and lead to more debts and reliance on exports and cash crops, whose effects on environmental degradation and overuse of agricultural land are apparent. NGOs are increasingly interested in sustainability and they realize that local development initiatives will be sustainable only when in partnership with a supportive national development system. That is why they are paying more attention to local and national governments which control the resources and policies that have a bearing on grassroots development, in the hope of influencing them in favour of an institutional and policy setting that is supportive of sustainable development policies. This task requires NGOs to have intra-and inter-organizational skills as well as technical competence and close collaboration with each other.

NGOs now see the future development of African countries as achievable through broader participation in the decision making process. They see such participation as informing both national and local development decisions. In conferences, NGOs have made it clear that the absence of full democratic rights in Africa is the main cause of Africa's economic decline. That is why they are advocating human centred development, the democratization of the development process, and are taking the lead in advocating policy and institutional reforms which are supportive of sustainable development and democracy and would stop the downward spiral of economic inequity and ecological destruction.

NGOs have identified the pressing problems facing African countries as being the undemocratic systems of government and the unequal, unsustainable, misguided and inappropriate development strategies. Consequently, they have proposed strategies by which equitable distribution of the benefits of sustainable development could be achieved; these include reforming of the world's trading system, more financial resources to African countries, and a reduction in military expenditure. They are critical of the institutions and policies causing environmental, economic and social degradation. For example, transnational corporations (TNCs) are under attack because of their role in the extraction of natural resources and despoliation of the environment, and international financial institutions are decried for opening up African economies for resource extraction by TNCs.

For NGOs to be able to promote democracy and sustainable development, they are recommended to "network; build up relationships with governments; participate in development planning committees at regional, provincial, district and ward levels; achieve economic self-reliance and mobilize, build up and expand constituencies," otherwise their democratizing potential will be limited. The existence of a dense network of autonomous grassroots development organizations and African NGOs on a substantial scale is essential in order to exert pressure in the interests of the poor and to act as their representatives, as well as a countervailing force to the power of the state, bureaucrats and local elites. NGOs also need to work to strengthen the organizational, technical and managerial capabilities of grassroots organizations so that the latter are able to stand up and press for demands and hold governments accountable for their actions.

By strengthening and developing grassroots organizations' capacities, NGOs will be enabling them to ensure resource management and control which correspond to the specific local context. Considering the climate in which NGOs must operate, their position on the African scene can sometimes be unenviable. NGOs have to contend with policy, economic and above all, political instability with all their ramifications. In some countries, the rise of political pluralism and the decline of centralized, one-party regimes has been replaced by fragmented special interests which have produced polarization, violence and political paralysis.

As if this were not enough, mutual suspicion and hostility sometimes characterize government-NGO relationships. A confrontational relationship with governments does little to enable NGOs go about their tasks effectively.

NGOs may face other threats, too, including disbanding or control. Other permanent threats to NGOs are administrative cooptation, appropriation, harassment and politically-motivated legislation. NGOs have also to contend with vested interests, which include bureaucrats, politicians and rural power elites, many of whom will oppose attempts to transfer power, responsibilities and resources to local institutions. Vested interests which have gained from non-democratic regimes are likely to oppose political and economic reforms. In such circumstances, governments, as well as vested interests, can constrain NGOs' ability to promote democracy and sustainable development. One way to overcome this is for NGOs to strive to create political space, build coalitions of friends and identify common points with governments, and then use this strength to chip away at the power of the vested interests.

Balance Sheet

The major sources of funding for NGOs continue to be voluntary private sources and governments. Where before African governments were the only major recipients of official aid from donor countries and multilateral institutions, nowadays more aid is being channelled through NGOs, many of which are perceived by the donor community to work more efficiently in participatory development and to operate in those areas which are not accessible to governments. The effect of this re-direction of some of the aid hitherto going to African governments has meant that these governments are being bypassed as implementing agents. This is leading to interesting examples of how governments are trying to 'coordinate' NGOs in the hope that they might be able to 'control' them now that they are recipients of more official funding.

While increasing availability of public funds for NGOs has been welcomed, for it expands their operations, NGOs are concerned that increasing acceptance of such funds could compromise their development goals, with the risk that they will be increasingly seen as agents of governments and multilateral institutions rather than as partners in development. The central challenge facing NGOs is how to maintain their voluntary character while becoming increasingly effective in their work.

NGOs realize that dependence on others for funding may compromise their flexibility to deal with pressing development issues. It may also undermine their ability to speak out against those policies of funders which they see as harmful to the interests of the poor. The other quandary which NGOs face is that while they advocate sustainable development and democracy for those they support, they themselves are, in many instances, neither democratic nor self-reliant.

In highlighting the potential and actual impact of NGOs as promoters of sustainable development, it is important that we do not lose sight of the difficulties NGOs face in achieving their goals. A recent series of studies by the UK's Overseas Development Institute (ODI) in Zimbabwe, India and Bangladesh suggests that NGOs have not been successful in (a) benefiting the poorest households, (b) benefiting women, and (c) ensuring self-sustainability of local NGOs. These findings serve to remind us that the tasks that NGOs set themselves are not easy to achieve and that a lot more will have to be done within and outside the NGOs if they are to be able to promote development that is balanced between local short-term use of man's environment and the maintenance and enhancement of long-term productivity.

NGOs have yet to make a significant impact both at national and local levels. In part, this inability stems from their lack of leverage. At the local level, they have yet to develop into an effective force which could counter the presence of local elites and influence decisively people's attitudes as well as local development policies. NGOs have many constituents, namely, donors, beneficiaries, policy makers and the public. These constituents have conflicting expectations. NGOs need to develop mechanisms for multiple accountabilities. Currently such mechanisms do not exist. This serves to constrain NGOs' effectiveness and blunts any claim to democratic credentials.

The political stance of many African NGOs contrasts with that of their Latin American counterparts where NGOs go as far as providing "trained personnel to occupy government positions." Many Latin American NGOs have had a history of directly and openly supporting social movements in opposition to military regimes. NGOs in Africa have not taken that high a political profile. For many African NGOs, politics is a forbidden and dangerous area. They argue that the best way for them to fulfil their development roles is to remain apolitical. Yet this is an unrealistic strategy, not least because many NGOs' operational activities are themselves political in that they seek to shift existing inequities in favour of the poor. Instead, they prefer to promote democracy by supporting grassroots organizations with resources, training and information. By their own actions, especially in participatory projects, NGOs have created practical schools of democracy from which members from grassroots organizations are able to follow democratic practices. Nevertheless, the feeling is that NGOs in Africa should do much more to contribute towards a more democratic Africa.

We have put forward a view that democracy and sustainable development are two sides of the same coin. The two concepts are underpinned by equality of access to resources, improvement in living conditions and commitment to democratic decision making. Both concepts, in practice, promote the idea of humanity being empowered and self-reliant, free and very much in charge of its destiny, in partnership with others.

The immense development challenges posed by Africa cannot find all the solutions in NGOs. NGOs have great capacity but cannot be the panacea to the challenges. That said, for NGOs' potential and impact to materialize fully, they must have more effective systems of internal monitoring and self-evaluation and must achieve self-sustainability. Somehow, they must acquire the political skills which will see them through the complex socioeconomic and political maze of the African situation.

Strategically, close collaboration with national governments is essential and, from time to time, NGOs must not hesitate to appeal to multilateral donors who have leverage and could act to protect them against political interference.

The ability of NGOs to promote democracy and sustainable development policies in Africa should be considered in the context of the states' preparedness to accept reforms which permit "devolution of power and responsibility for resource use and management from the centre to the communities."

The goodwill of national governments, as well as of the international community, is crucial to NGOs' successes or failures, not least because, in the case of the latter, they have the resources to create a supportive international climate that could reverse the flow of resources from African to northern countries and support better terms of trade and accessible markets.

As for national governments, they should show goodwill by creating internal conditions which permit a liberal democratic framework as well as a supportive legal framework.

Both macro-and micro-economic policies which are favourable to NGOs would offer a conducive framework within which NGOs would flourish. In their quest for sustainable development and a democratic society, therefore, NGOs must strive to attain this goodwill. It underlies all else.

PEOPLE'S ORGANIZATIONS AND DEVELOPMENT NGOS

A close conceptual and operational link exists between people's participation and people's organizations. Active participation of rural people can only be brought about through local community and membership-based self-help organizations whose primary aim is the pursuit of their members' social or economic objectives.

People's organizations are voluntary, autonomous and democratically controlled institutions including traditional community councils, informal groups, cooperatives, rural workers' organizations and peasant unions, women's associations, etc.

Some local people's organizations may establish higher-level federations at provincial, national or international level in order to increase their self-help capacities and bargaining power, and to promote participatory development at local level. However, the vast majority of the rural population are still not organized in groups and are therefore not benefiting from the dynamics of such groups.

Participation through people's organizations is enhanced at local level through the work of development NGOs that aim at improving the social and economic conditions of rural people, especially the poor.

Some development NGOs are membership-based, accountable to local associations which establish them, but the majority are not. The support they provide to grassroots groups takes various forms: training, technical support, research, assistance in formulating projects, exchange of information and experiences.

NGO approaches to participation, geared to enhancing the self-reliance of people's organizations, are increasingly relevant when structural adjustment measures are obliging governments to cut back on state services. They help people's organizations to build up a substantive platform of awareness and initiatives on the basis of which they can participate meaningfully in planning and implementing government-promoted development programmes.

Over the past few years most governmental and multilateral development cooperation agencies have made serious efforts to strengthen their collaboration with the non-governmental sector, due in large part to a recognition of the relevance of NGO experience and the importance of their grassroots outreach. FAO established an NGO programme over 30 years ago, the Freedom from Hunger Campaign/Action for Development (FFHC/AD).

Already in 1971 promoting people's participation in development was one of the main themes debated by NGOs associated with FAO at the Fifth FFHC/AD Conference.

Over the intervening years, FFHC/AD has built up continuous working relations with NGOs and NGO networks in the South and a programme of support for their initiatives. At the same time, most FAO technical divisions and country offices have initiated collaboration with NGOs relevant to their areas of work. The establishment of World Food Day has given FAO an additional promotional instrument to do so.

BACKGROUND AND STRATEGY OF THE PLAN OF ACTION

In 1989, following requests from Member Governments and the Rome-based Group of International NGOs, the Committee on Agriculture (COAG) and the Council examined the issue of people's participation and its implication for FAO's activities in rural development. They recommended that the concept of participatory development be integrated into all development policies and programmes of FAO and also suggested that FAO develop a Plan of Action for People's Participation for discussion in due course by the FAO Council and Conference.

A Plan of Action for People's Participation was therefore submitted to the 99th Session of the FAO Council, which broadly endorsed its objectives and the areas of action identified in it. The Council agreed that, in general, the Plan of Action constituted a suitable framework for future action and provided useful guidelines for interested governments to facilitate the process of people's participation.

It stressed that the Plan of Action should fully recognize and respect the sovereignty of Member Governments. The view was also expressed that the Plan should be further elaborated and translated into concrete activities. The Council agreed that the Plan of Action should be revised taking into account the views expressed during the Council discussion and be submitted to the Twenty-sixth Session of the Conference for consideration and adoption.

Accordingly a revised Plan of Action for People's Participation is submitted in the following paragraphs for consideration.

The overall aim of the Plan is to ensure active participation of people in the achievement of sustainable rural development.

While it is recognized that other factors relating to social, economic/ financial and technical aspects do play an important role in achieving this objective, the active participation of rural people, including disadvantaged groups, acting through voluntary, self-reliant organizations of their own choice is equally important.

Without such participation, rural development initiatives are unlikely to be sustainable in the long run and rural inequities are unlikely to be redressed. It should be recognized that the policy decision and responsibility for formulating and implementing participatory development policies and approaches rest with the governments of individual member countries.

Hence, the policies and programmes for people's participation envisaged in the follow-up to the Plan of Action must fully recognize and respect the sovereignty of Member Governments.

The promotion of people's participation in rural development can be operationalized if certain conditions are met, namely:

- that general public awareness exists of the benefits of using participatory approaches to rural development, and that key policy and decision-makers within government and other development agencies take action in this area:
- that the legal and policy environment existing within the country is

favourable to the participatory, collective self-help initiatives of rural people, especially the disadvantaged; and

- that the self-management capacities of rural people's organizations are sufficient to sustain such initiatives.

Policy advice and technical assistance to Member Governments aimed at facilitating the decentralization of public administration decision-making will be critically important in the creation of more favourable legal and policy conditions for people's participation.

The development of new, more flexible and participatory administrative and operational procedures and methodologies for implementing participatory rural development actions and supporting local organizations will likewise be important.

Furthermore, the design and implementation of new tools and mechanisms for monitoring and evaluating people's participation and for improving dialogue among people's organizations, government, NGOs and other development agencies on people's participation issues will be essential in building greater public awareness and understanding of the benefits of participatory rural development approaches.

In order to achieve the goal of sustainable rural development with equity through people's participation, the Plan proposes that action be taken in the following seven areas:

- Promotion of greater public awareness of the role of people's participation and people's organizations in agricultural and rural development;
- Creation of favourable legal and policy framework for people's participation;
- Strengthening internal capacities of the rural people's organizations at local and national levels;
- Decentralization of government decision-making;
- Promotion of increased dialogue and technical collaboration between governments, development agencies and people's organizations;
- Introduction of appropriate operational procedures and methods to facilitate wider participation;
- Monitoring and evaluation of people's participation.

Action proposed in each of these areas is outlined in the following sections. Each section begins with a summary description of central issues and problems, followed by specific recommendations for consideration by Member Governments and FAO.

Action Areas

Promotion of greater public awareness of the role of people's participation and people's organizations in agricultural and rural development

Issues: Achieving sustainable rural development and environmental protection requires the cooperation of large, sometimes diverse populations residing within specific ecological areas. Such cooperation can be enhanced through the voluntary and cooperative action and participation of all sectors of the rural population, including women and other disadvantaged groups. Full recognition of the inherent advantages of people's participation in mobilizing rural communities for sustainable agriculture and rural development is an essential first step. Many key decision-makers need to be informed about and convinced of the intrinsic benefits that flow from enhanced people's participation in development programmes and projects.

Proposals for Action

For Governments:

- Create greater awareness among government officials of the benefits of adopting participatory approaches in order to reach rural people.
- Establish clear government policies and regulations for the training of government officials in the principles, practice and benefits of people's participation in development.
- Apply communication methods and materials, to be used by government staff and people's organizations, in the promotion of people's participation and in the sharing of knowledge and skills.
- In order to reach all sectors of the rural population utilize a variety of organizational modalities, such as small informal groups, traditional community associations, cooperatives, unions, etc.
- Mobilize rural communities to achieve sustainable rural development objectives using participatory approaches, acting through existing or through new rural people's organizations.
- Ensure that mobilization activities are focused on satisfying community needs and produce tangible benefits for those involved.

For FAO:

- Conduct case studies and research on the effectiveness of participatory approaches and people's organizations in mobilizing rural communities for sustainable agricultural and rural development and environmental protection.
- Assist governments in testing the effectiveness of different organizational methodologies for mobilizing rural communities,

satisfying community-identified needs and producing tangible benefits for those involved.

- Collect and disseminate to Member Governments, NGOs, other people's organizations and the general public information materials on examples of people's participation activities, approaches adopted and performance and benefits.
- Develop informational training materials on participatory rural development project design, monitoring and evaluation in order to enhance among concerned FAO technical staff, the awareness of people's participation issues and to assist in the integration of people's participation elements into FAO programmes and projects.
- Assist governments in developing new training approaches aimed at sensitizing key government and development agency decision-makers to the merits and value of using more participatory approaches in rural development.
- Through the World Food Day Network, mobilize governments and NGOs for the creation of a positive attitude towards people's participation.

4

Challenges for Rural and Social Development

Some of the important social problems like poverty, ignorance, over-population and rural backwardness are of a general nature and, in varying degree, they are influenced by factors like squalor and bad housing, malnutrition and physical and mental ill-health, neglected childhood, family disorganization and a low standard of living.

For along time, society has remained apathetic to these conditions, but with the awakening of political consciousness and the enthusiasm of organizations and workers to improve social conditions, there is a possibility of developing programmes which could gradually remedy the present situation. The economic programmes of the Five Year Plan will mitigate these problems to some extent, but the gains of economic development have to be maintained and consolidated by well-conceived and organized social welfare programmes spread over the entire country. It is proposed to consider some of the more important problems of social welfare which need the special attention of both State and private welfare agencies.

The principal social welfare problems relate to women, children, youth, the family, under-privileged groups and social service. The social health of any community will depend a great deal upon the status, functions and responsibilities of the woman in the family and in the community. Social conditions should give to the woman opportunities for creative self-expression, so that she can make her full contribution towards the economic and social life of the community. Problems relating to health, maternity and child welfare, education and employment.

Some problems of women have to be dealt through social legislation, but other problems pertaining to health, social education, vocational training, and increased participation in social and cultural life, provision of shelter, and assistance to the handicapped or maladjusted call for programmes at the community level. As women have to fulfil heavy domestic and economic responsibilities, adequate attention has to be paid to the need for relaxation and recreation both in the homes as well as in the community. The welfare agencies have catered to some extent to the needs of the widow and the destitute woman, but the quality of the service rendered by them and the nature of their work needs to be surveyed.

Considering the numbers involved, the needs of children should receive much greater consideration than is commonly given to them. There is a growing demand for child health services and educational facilities. The standard of child welfare services in the country can be improved if the rate of increase in population is reduced. Problems relates to family planning, children's health, infant mortality, education, training and development have been discussed elsewhere in this report. Malnutrition is perhaps the major cause of ill-health and lack of proper growth of the child. The feeding of the child in the early years is the responsibility of the family, and is dependent upon economic conditions and traditional food habits. The nature and extent of malnutrition has to be determined, and resources have to be found to supplement and improve the diet of children through schools and community and child welfare agencies.

The problem of children's recreation and development outside educational institutions has received some attention during recent years, but play activities of children are considerably restricted in urban areas on account of the environmental conditions, lack of adequate space, and, to some extent, neglect of this vital need of the child by the family and the community. Not enough is known about the work of private agencies for the welfare of destitute and homeless children. The juvenile courts and children's aid societies have so far touched a fringe of the problem of children's welfare. Certain special aspects may be briefly mentioned. The existing facilities for handicapped and deficient children are far from adequate and suitable agencies have to be created. Hospitals provide treatment for polio, congenital deformities, fractures, bone disorders and other diseases, but there is a need to extend existing services and provide special institutions and care for disabled and crippled children.

At present deficient children attend educational institutions together with normal children and seldom receive treatment and special training to enable them to overcome their handicaps. The subject needs to be studied carefully. The problem of juvenile delinquency has already received considerable attention and many of the States have special legislation. Juvenile delinquency may often be the result of poverty and many offences may be traced to the connivance or support of adults. The youth constitute the most vital section of the community. In recent years, young people have had to face and have been increasingly conscious of problems such as inadequate educational facilities, unemployment, and lack of opportunity for social development, national service and leadership. The problems of health, education and employment of youth have been considered as aspects of national problems in these fields.

Social welfare is primarily concerned with the improvement of services provided for the benefit of youth by welfare agencies with the object of promoting development of character and training for citizenship and for physical, intellectual and moral fitness. It is necessary to encourage initiative among youth so that through their own organizations, they can develop programmes of youth welfare and national service. Ways must also be found to give opportunities to youth for active participation in constructive activity. Such training and experience will equip them for shouldering the responsibilities of leadership in different spheres of national life.

Traditionally, the family has been left largely to its own resources to deal with most of its problems, although in some cases it may be assisted by the larger community groups (such as caste) to which a family may belong. General problems relates to health, education and employment. Questions relating to status and rights, property, inheritance, etc., are the subject of social legislation.

The gradual break-up of the joint family and the emergence of the small family have increased its economic problems and burdens. Family responsibilities have now to be borne at a comparatively younger age by the head of the small family than happened in the joint family. This creates the need for greater guidance and assistance in dealing with family problems. The increasing complexity of the social situation and handicaps arising from physical disability, ailment or unemployment render it more difficult for the family to provide a sense of security to its members.

This fact suggests a number of problems which, along with other problems such as divorce, desertion, and treatment of mal-adjusted members of the family, need to be studied carefully if welfare agencies are to develop suitable methods of treatment for guiding and assisting those in need.

There are a number of under-privileged communities such as the scheduled tribes, scheduled castes and other backward classes including criminal tribes. The problems of poverty, ill-health, and lack of opportunities for development affect them to a larger extent than many other sections of the society. The main problems to be considered under the description of social vice are prostitution, crime and delinquency, alcoholism, gambling and beggary. These problems have existed for a long period, although necessarily their nature and extent vary according to the prevailing social and economic conditions. Some of them have to be dealt with largely by local communities, and the approach and treatment have to be varied from place to place.

The character and magnitude of these problems of social defense have to be determined carefully before the value and efficacy of the existing agencies and programmes could be assessed. Social legislation deals with many of the social evils with a view to controlling and even eradicating them, but its actual implementation needs to be watched.

Among the practical problems to be resolved are the demarcations of the relative roles of State and private agencies, determination of the machinery of enforcement, estimation of the resources required, examination of methods, development of correct programmes, and creation of public opinion in favour of an objective and dispassionate approach to the problems of social vice.

As the social structure becomes more complex, the State is called upon to play an increasing role in providing services for the welfare of the people.

The Central Government, the various State Governments and local self-governing bodies, each in its own sphere, have to ensure that they have at least the minimum administrative machinery for dealing with social "problems. What form this machinery takes will depend on their particular circumstances and requirements, but it is certain that without the necessary machinery they will not be able to pursue their programmes.

THE CHALLENGES OF PARTICIPATORY RURAL DEVELOPMENT

Despite our best efforts, perfect, transformative, empowering participation is probably unattainable most of the time, except perhaps where it was not necessary to begin with. The question then is whether we can have a 'good enough participation', what that would look like and what it would require. Perhaps the main criterion would be that the interventions associated with it should do no harm, and preferably some good, in terms of the main objective of poverty alleviation, while at the same time contributing towards creating at least some of the conditions that enable participation, agency and the full exercise of citizenship by the poor to take place. If such interventions remove some of the constraints of everyday life, by increasing the productivity of agricultural labour or bringing clean water closer to where people live, alleviating the burden of disease caused by malnutrition and lack of hygiene, decreasing vulnerability to natural disasters or to the vagaries of climate and rainfall, or making people less dependent on relations of patronage for daily subsistence, they are a step in the right direction (Cleaver, 2004). If this is done for instance through the permanent and expanding presence of an organization that offers 'better technology and more affordable inputs rather than autonomy and independence' (Mosse, 2001: 34), it may not fit in with the fashion for empowerment and farmer-managed development, but may still represent significant progress.

Multidimensional Rural Poverty

The complexity and multi-dimensionality of poverty in general, and rural poverty in particular, are generally accepted nowadays. A recent ESCAP document calls for: ... a deeper knowledge and understanding of the emerging dynamics and local processes that cause and maintain rural poverty. The rural poor face multiple deprivations from lack of assets, isolation, alienation, dependence, powerlessness, vulnerability and lack of freedom of choice.

Disparities also exist within rural areas: in particular, disparities between ruling rural elites and small farmers or landless households; and disparities among farmers over access and rights to fertile lands. In this context, a distinction between the different but closely related dimensions of poverty is important: (a) a lack of income and productive assets; (b) a lack of access to essential economic and social services; and (c) a lack of power, participation and respect. These different dimensions of poverty reinforce each other, keeping the poor trapped in poverty (ESCAP, 2007: 8).

Lack of awareness of this multi-dimensionality, or a reluctance to confront it, may explain much the failure of past rural development strategies in reducing rural poverty. The implication is clear: new, sharper, multi-targeted rural development strategies are called for, taking this complexity and multi-dimensionality fully into account and going beyond the old approaches that were based solely on the growth of agricultural productivity and focused almost exclusively on the small farmer.

An approach to rural poverty and rural development that appears capable to encompass this complexity and multi-dimensionality is the so-called sustainable livelihoods approach. As Ellis and Biggs (2001: 445) emphasize, this approach '... embodies no prior requirement for the poor rural individual or family to be a "small farmer". The livelihoods approach takes an open-ended view of the combination of assets and activities that turn out to constitute a viable livelihood strategy for the rural family'. The same authors add that empirical research indicates that farming amounts on average to only about 40-60% of the livelihood 'package' put together by rural households in South Asia and sub-Saharan Africa11, and conclude:

If a new paradigm of rural development is to emerge, it will be one in which agriculture takes its place along with a host of other actual and potential rural and non-rural activities that are important to the construction of viable rural livelihoods, without undue preference being given to farming as the unique solution to rural poverty. It is in this sense that the cross-sectoral and multi-occupational diversity of rural livelihoods may need to become the cornerstone of rural development policy if efforts to reduce rural poverty are to be effective in the future.

Such paradigm will probably combine participatory approaches with rural livelihoods, given their compatibility and potential for mutual reinforcement.

Contextual Factors and the Role of the State

The success or failure of participatory rural development interventions depends as much on the conditions surrounding the particular intervention as on the quality of the work done at the local level. Among the conditions most frequently mentioned are economic growth, a favourable political and administrative environment characterized by administrative decentralization and more generally good governance at the macro level, and the presence of cultural traits or ideologies that favour participation, self-reliance and collective action by the poor.

It must, however, be kept in mind that (i) as experience has repeatedly shown, there is no automatic cause and effect relationship between economic growth and poverty reduction; (ii) decentralization and good governance work primarily by removing constraints and barriers; and (iii) even when culture and ideology favour collective action by the poor, there are limits to what they can do, as their agency is constrained as much as it is enabled by structure.

Although the poor can be resourceful and imaginative, they are unlikely to be able to escape poverty by themselves. This means that neo-liberal solutions to the problem of poverty are unlikely to work: the problem will just not go away without deliberate and proactive government policies, backed by appropriate resource allocation. This does not imply a return to the government-led, topdown interventions of old, which did not work either: rather, the state must work with the poor, creating the conditions for their participation and supporting it with resource transfers, stimulating pro-poor growth and working to reduce inequality and promote social justice.

Policies specifically aimed at participatory rural development include legislation for rural people's organizations, including full freedom of association; reorientation of the delivery system, in particular the extension services, towards the needs of the rural poor; full integration of women in development; decentralization of decision-making, planning and resource allocation; rural poororiented research, extension, input supply, credit and marketing, supported by the necessary financial resources and aimed at enhancing income-generating activities, both agricultural and nonagricultural; and just fiscal and pricing systems (van Heck, 2003). Participatory rural development projects must involve government institutions, such as line departments or local governments, banks, training and research institutions, women and youth councils; they must also continue to involve civil society organizations at all levels.

Government institutions need to become more open to the advantages of using local knowledge for improved productivity in farming, water conservation and natural resource management, and of associating local communities and small farmers to their research. Equally needed is research on sustainable and labour-intensive agricultural production systems, rural off-farm employment and enterprise development (Polman, 2001: 10).

Pilots, Innovations and Scaling Up

Many of the most innovative rural development projects start small. This makes it possible to experiment with new approaches (e.g. to participation) or technologies, while keeping costs and risks within acceptable limits. Such projects are thus, for all practical purposes, pilot projects. This means that, if they are successful, they will at some point need to be scaled up. Scaling up has its own problems that need to be resolved before successful projects– or parts of projects – can yield significant benefits. These problems are not trivial, and if they are not resolved, the initial promise may lead to disappointing results.

On the basis of a number of case studies of scaling up of technical or institutional innovations, Biggs (2006: 16-17) points out in relation to poverty reduction and social inclusion that effective innovation in the policy and institutional arenas is generally location and time specific, and *has* to be, because the significance of institutional innovations can only be understood with reference to time, place, culture and political context. In the cases studied, ...major and significant institutional innovations were coming up all the time from within the local context. Informal, unplanned R&D was always taking place [...] In addition, and significantly, in 'scaling up' processes, it was the local institutional innovations that were the critical elements that gave rise to the way the technology or institutional model was scaled up and to the pattern of outcomes as regards who benefited from the process. Consequently, for those concerned with being effective in poverty reduction and social inclusion, the most important research starts at the so-called 'scaling up' stage.

Scaling up is not a minor 'development' job to be done at the end of an R&D pipeline [...] Institutional innovation in the local context is not in any way similar to the 'adaptation' to local circumstances of outside technical innovations.

Biggs then goes on to argue for the importance of 'learning from the positive'; policy and development personnel should constantly be on the look out for the positive changes that are always taking place wherever poverty and social exclusion exist, and monitor, learn from, support and promote these changes.

Scaling up can be interpreted in two separate and complementary ways: (i) as replication on a large scale, from a small to a large project (a variant of which is from a small to many coordinated small scale projects in different places); and (ii) as 'mainstreaming into policy making'. Each of these meanings implies different challenges. Large-scale replication of a small participatory rural development project requires that many facilitators or extension agents pay the same amount of concentrated attention to their work and their clients as did the initial pilot team. They have the advantage of being able to learn from other people's mistakes; on the other hand, what was for the initial participants a process of learning by doing, with a significant research component, is now something to be taught and learnt, then applied. There is less incentive to perform well, and it is easier to fall into routinized, mechanical approaches. To deal with these challenges, at the very least motivation, indoctrination and institution building (including the provision of suitable incentives) are needed before the replication starts, close monitoring and control at the beginning of implementation, and spot checks throughout.

As to scaling up in the sense of mainstreaming into policy making, its success is a function of the capacity of institutions and organizations to learn, and of the existence of enabling environments that provide the conditions, stimuli and incentives for institutions and individuals to act in desirable ways. Focusing on the scaling up of participatory watershed management in India, Kolavalli and Kerr (2002) maintain that, given the limited availability of NGOs, a realistic strategy must be to improve the capabilities and incentives of government agencies to work more like NGOs. They argue that institutionalizing participatory approaches in government bureaucracies may include (i) creating an enabling environment where the necessary conditions and appropriate incentives exist, plus (ii) strengthening upward accountability and (iii) introducing downward accountability. Joshi and Moore (2002) argue that the outcomes of public programmes to tackle poverty in developing countries are more likely to be decided in the implementation than the decision-making phases of policy making, and conclude that it is vital to try to 'plug the gaps' at this stage.

Searching for ways in which this might be done through a 'self-correcting' mechanism that creates an enabling environment for the poor, they focus on the extent to which anti-poverty programmes succeed in mobilizing their own beneficiaries. They argue that the likelihood of collective action by the poor to defend their rights in the context of anti-poverty interventions depends on the *predictability* of those interventions from the point of view of the poor, or of the social activists and political entrepreneurs needed to galvanize them into collective action.

This in turn has three components: *credibility* of the officials implementing the anti-poverty programmes; *stability of the programmes* over time, in content, form, procedural requirements, etc.; and *formal entitlement,* which includes but is not limited to legal rights to benefits of the programme. Such predictability, argue these authors, encourages and supports collective action in favour of the anti-poverty programme.

The Importance of Territory

From the local to the territorial it is a small step, but one that adds a rich layer of meaning to the discussion. The territorial dimension can contribute significantly to participatory rural development, by providing it with a spatial and social context within which participation may become easier and more meaningful. In recent years, the concepts of *territory* and of *territorial competitiveness* have acquired considerable importance as a result of economic and social changes within countries, in the context of accelerating globalization. The word 'territory' designates a rural region, which will normally include one or more small or intermediate urban centres together with their rural hinterland.

Because of its scale and of the closeness of the interactions that take place within it, the territory is increasingly seen as a privileged space for participation in rural development. The concept of territorial competitiveness starts from the usual definition of the term competitive and adds to the concept a number of normative elements that broaden its meaning and enrich it as a policy objective. Interpreted in this broader way, "a territory becomes competitive if it is able to face up to market competition whilst at the same time ensuring environmental, social and cultural sustainability, based on the dual approach of networking and inter-territorial relationships".

In order to strive for territorial competitiveness, the actors in the territory engage into negotiation leading to the formulation of a collective *territorial project:* a vision of the future of the territory in which territorial competitiveness is assured and therefore the territory fulfils its potential. This depends on local actors and institutions acquiring four types of skills: the skills to take joint action, to assess their environment, to create links between sectors by ensuring that maximum added value is retained, and to liaise with other areas and the rest of the world (*op.cit.*). These skills link to the so-called "four dimensions" of territorial competitiveness, present in each area in different combinations: social, environmental and economic competitiveness, plus positioning in the global context. These processes are seen as forming part of a long-term approach that is at the core of each area's development strategy (Farrel and Thirion, 1999). The presence of divergent or even opposed interests implies that negotiation, usually managed negotiation, is explicitly present and plays a key role in the definition of the territorial project, itself a process of social learning. The presence of an urban centre (or more), and its linkages with the rural areas, is a crucial part of each territory's identity, a focus of negotiation and participation processes and of local institutional innovation, and a key factor of success in achieving territorial competitiveness. In order to move from an analysis of the situation to the identification and design of a territorial project, to charting out a path for the territory's future, it is necessary to carry out an analysis of what is known as *territorial capital,* defined as: ...all of the elements available to the area, both tangible and intangible, which in some respects constitute assets and in others constraints.

SOCIO-CULTURAL AND RELIGIOUS FACTORS

The socio-cultural and the religious factors refer to the social customs, cultural traditions and religious faiths which play dominant role in the life of people especially in poor and backward communities. The social factors induce a person to have more sons as the sons are believed to be the reliable means of social security and social status for a family. Moreover, they are deemed to care their old, physically wasted and worn-out parents, on one hand, and to provide safety to the family in case of conflicts and death of supportive member/s of the family. The religious factors refer mainly to four beliefs. One states that peace is rendered to the soul after death only if the cremation is performed by a real son. The second speaks of higher spiritual gain or place in Elysium of heaven after death for a person having more sons. The third relates the production of more children to religious service and the fourth specifies child bearing as the pious duty of a woman because she is believed to be sent by God only for increasing progeny.

ECONOMIC FACTORS

Apart from the socio-cultural and the religious factors, the economic factors are very much effective like the biological factor. A person in acute economic privation leaves even deeply instilled socio-cultural, religious and some times biological allurements too for economic gains. The need of economic support in old age and the poverty are two main economic factors making people tend to produce more children. In old age, when a person becomes generally sick, pulled, worn-out, and non-earning, he becomes dependant to his heirs for food, clothing, medical treatment and other expenses. More the sons or heirs he has, lesser will be the share of burden of supporting him. As regards to poverty, it generates three causes of inducement to the high rate of child-birth.

(a) The prevailing media based fascinating means of entertainment are not only out of the reach of poor man but these have also snatches the old socio-cultural, cheap and some times free entertainment sources from him. Therefore a poverty stricken person has to search the way of entertainment in sex and that too being circumstantially unprotected whereby childbirth goes on taking place one after one successively.

(b) In poor families a child becomes earning hand at the age of seven or eight yrs. On account of the subsistence level of family's living, the expenditure on child's feeding is considerably lower than the wage he earns. Therefore, the surplus of his earning over his consumption adds to family income and thus contributes to the uplift of family's standard of living. The sentiments and feelings regarding education or future welfare of the child droop before the agony of unsatisfied basic needs due to privation. Therefore, a child in a poor family is proved an asset rather than liability, in its stead.

(c) There are some types of family occupations where a number of faithful workmen are required. The required manpower from own family is most desirable there. If a person is owner of a series of units of small scale industries, cottage industries, small business units etc. the hired managing persons generally prove costlier, unfaithful and non-devoted. If a family member is deployed at each such unit the safety and profitability is increased. Similar is the position in a single cottage industry unit where margin of profit is low and hired labour makes the profit uncertain. If family manpower instead of hired one is used the profitability there becomes increased on account of devotedness, faithfulness and per need flexibility in working hours of the working force. Moreover, wage rate of a family labour is generally lower than that of a hired labour because the cost of living of one member of the family is lower than that of the whole family of hired man. Therefore the requirement of man power is tried to meet out by producing more children in the family.

ORGANISATIONAL STRUCTURE OF THE DRDAS

Each district will have its own District Rural Development Agency. Ordinarily it would be a society registered under a Societies Registration Act.

In respect of such states where DRDA does not have a separate identity a separate cell should be created in Zill Parishad which maintains a separate identity and separate accounts, so that the accounts are capable of being audited separately. This Cell should be directly under the charge of CEO or alternatively an officer who has the qualifications to be a Project Director. The accounts of DRDAs should under no circumstances be merged or amalgamated with those of the Zilla Parishads.

In order to be effective, the DRDA must have an appropriate staffing structure as well as suitable personnel policy. Keeping in view the rule and functions of the DRDA, the staffing structure of DRDA must include positions for Planning for poverty alleviation, Project formulation, Social Organization and Capacity builing , Gender concerns, Engineering Supervision and Quality control, Project Monitoring, Accountancy and Audit functions as well as Evaluation and impact studies.

An indicative staffing structure is given in Annexure and is also detailed below.

The State Government may modify the structure suitably, but without altering the basic design, to take care of the needs of individual districts keeping in view their size as well as specificity. However, this will also be subject to the overall ceiling of administrative costs admissible to the DRDAs in the State. By and large the staff appointed should be dedicated to DRDA-related works and should not be frequently transferred.

The Staffing structure will as follows:

Project Director

Each DRDA should be headed by Project Director, who should be the rank of a Additional District Magistrate. The Project Director should preferably be a senior scale officer of All India Services or a senior officer of the State Service, eligible of appointment to the All India Services. He would be in overall charge of the activities of the DRDA and responsible for interaction with the District/State administration as well as with the Government of India. The PD should be exclusively for DRDA work.

In some State, such as Maharashtra, the CEO of Zilla Parishad is the Chairman of the Zilla Parishad. Government of India have suggested to all the State Government that the chairman, Zilla Parishad should be the chairman of the governing body of the DRDA. In the light of this, such states, the CEO of Zilla Parishad could also be designated as the Project Director of the DRDA.

Each DRDA should have the following wings:

(a) Self - employment Wing;
(b) Women's wing;
(c) Wage employment wing;
(d) Engineering wing;
(e) Account wing;
(f) Monitoring and Evaluation wing;
(g) General Administration wing;

Self- Employment Wing

(i) The self-employment Wing, shall be headed by a Project Officer and should have APOs in the field of planning, social mobilization, credit and Technology. The Project Officer may be specifically responsible for any of these four functions. The APO (Planning) would look after the activity clusters, district/block. Village group plan, guiding the BDOs and others in plan preparation, planning for infrastructure including marketing infrastructure. He would coordinate with the district officers, the banks as well as other institutions in the district. The APO (Social Mobilisation) would look after group formation, capacity building, monitoring of groups, choice of activities for groups, release of revolving funds and coordination. The APO (Credit), who should be from the commercial banking sector, would coordinate with the banks in all matters relating to credit, including the interface between the bankers and the beneficiaries/ beneficiary groups, loan disbursements as well as loan recovery. The APO (Technology) would look after issues concerning Technology upgradation as well as transfer of technology.

(ii) The DRDAs would be drawing up projects for specific activities under the Self-employment programmes, it would be necessary to have suitable experts to oversee the successful implementation of such projects. As part of the overall programme management, the DRDAs may take outside expert on a consultancy basis. Secretary, Rural Development of each state should develop guidelines for selection of consultants so as to avoid any wrong use of such provisions and to see that only those who have a proven practical expertise are engaged. Such experts to be engaged on project specific basis will function under the overall control and supervision of the Project Officer, self-employment programms.

Women's Wing

In order to ensure that women receive adequate attention in all the anti-poverty programmes, a Women's Cell should be sit up in each DRDA. This cell would establish necessaty synergy with departments such as Women & Child Development, Education and health Departments to ensure that women not only receive their due share in the anti-poverty programmes but are also able to receive benefits of other programmes. The Women's wing will be headed by an Assistant Project Officer, who will function under the overall co-ordination of the Project Officer of the self employment wing. Assistant

Wage Employment Wing

The central concern of the DRDA in the wage-employment programmes should relate to planning, monitoring and vigilance by a technical wing. The DRDA should not concern itself with the actual implementation and execution of works which can be done by the line department / engineers or the Panchayati Raj Institutions. At the same time, the magnitude of the wage employment wing, with a Project officer assisted by a small complement of staff.

Watershed Wing

A Watershed Wing will be set up in the DRDA in all such districts where IWDP / DPAP /DDP is in operation. This wing should consist of a Project Officer, assisted by a small complement of staff. This staff would be independent of the programme support in the form of PIAs or Watershed committees.

Engineering Wing

Each DRDA should have a technical wing, which should also be responsible for innovations in design or use of materials, as well as training of the engineering personnel of the line departments or the PRIs.

This wing should be manned by an Executive Engineer assisted by one or two Assistant / Junior Engineers.

Account Wing

(i) Wherever it has not been done, the DRDAs should shift over to commercial accounting systems from the existing government account system. They should publish an annual report alongwith the balance sheet. The accounts wing of the DRDA should be headed by a Sr. Accounts Officer, either on deputation or by engaging the services of a chartered accountant. He should be supported by an Accounts Officer each for self-employment programmes and wage employment programmes duly assisted by accountants. Wherever the Watershed programmes (IWDP/DPAP/DDP) are under implementation, an additional post of Accounts Officer may be sanctioned. For Indira Awaas Yojana, one Accountant at the district level should be available to monitor the progress of the programme and the accounts.

(ii) One of the Accounts Officers should perform the role of internal audit.

Monitoring Wing

These should be a separate Monitoring and Evaluation wing headed by a Project Economist and functioning directly under the supervision of the Project Director.

Apart from monitoring the progress of all the programmes., this wing should also carry out evaluation /impact studies regularly by independent by independent institutions /experts including NGOs.

The cost of such studies will be met from the respective programme funds. This wing should also monitor issues relevant to poverty in the district.

ADMINISTRATIVE COSTS

Administrative costs were earlier admissible under different programmes. In order to enable the DRDAs to develop a proper personnel structure and to ensure its proper growth over years, the practive of meeting the administrative costs as percentages of each programme allocation has been given up. It will, on the other hand, be taken up as a new scheme called 'DRDA Administration with effect from 1st April, 1999. This scheme will be funded on a 75.25 by the central and state government.

Till 31.3.1999, administrative costs of the DRDAs were being met by way of a part of the programme fund of IRDP, JRY, EAS, DPAP, etc. With the introduction of the scheme of 'DRDA Administration' the administrative costs earlier available under different programmes stand withdrawn from 1.4.1999 and provision of administrative support to the DRDAs will be only under the scheme of 'DRDA Administration'.

Since the salary structure in different States different, the States may follow their own salary structure. However, the administrative cost ceiling fixed should be met entirely by the State government.

A maximum of 30% of the salary cost may be allocated towards contingencies inclusive of Rents, POL, office expenses, etc.

The administrative cost per district is fixed as follows:

- Category A districts (<6 blocks) Rs. 46 lakh per year
- Category B districts (6-10 blocks) Rs. 57 lakh per year
- Category C districts (11-15 blocks) Rs. 65 lakh per year
- Category D districts (>15 blocks) Rs. 67 lakh per year

The above limits will be applicable from the year 1999-2000. This ceiling will be raised every year, on a compounding basis, upto 5% to set off the increases due to inflation etc. The state governments are allowed a sum upto 10% of the above costs, to be calculated at 10% of the total admissible cost to all the districts in the state.

PARTICIPATORY PLANNING SYSTEM

A good way of examining the working of the set up created by these Constitutional Amendments is to examine the actual functioning of the decentralized system. We do so with three case studies of rural roads planning and their implementation. The first is from a developed district, Ahmedabad in Gujarat – a state with a long tradition of decentralized development and a reputation for some institutional reform. The second is from an area which has a long tradition of rural development, but not through government efforts. India's first cooperative sugar factory was set up at Loni, a backward area of western Maharashtra in Ahmednagar district. Its founders were the rural development leader, Vikhe Patil and the economist, D.R. Gadgil. The issue here concerns the coordination between people's representative institutions and a non-governmental organization (NGO) such as an all-purpose development cooperative. The third case study is from a backward area of a state, the Aurangabad Division of Maharashtra, which has a well-developed system of decentralized institutions but development planning at the local level has not yet been activated. This would be the case of the "problem area".

Ahmedabad Rural Road Development Plan

The Ahmedabad district is in central Gujarat and is dominated by the industrial town of Ahmedabad, where the first textile factory in India was set up in 1868 AD. In 1991, the district had a population of 4.80 million, of which 74.7 per cent was urban. The district consists predominantly of alluvial soil and has 90 cm of annual rainfall on average. The River Sabarmati traverses the district from a north-easterly to a south-westerly direction and ends in the marshes of the Gulf of Cambay off Dholera. The Khari, Meshwo, Onkar, Bhadar, Neelka and Utavali drain into the Sabarmati at Voutha. Rainfall is mostly from June through September and the rivers flood in this period. The district consists of the Talukas of Dascroi, Sanand, Dehgam, Dholka, Viramgam and Dhanduka.

A district road plan was prepared in the early 1970s when India's Fifth Five-year Plan laid down the objective of connecting of each and every village as a minimum need. This was approved later by the District Planning Board. First priority was given to villages with a population of 1,000 and above. The remaining villages were to be connected later. The plan consisted of works to connect every village to the nearest road. Villagers were not involved in the choice of the network although representatives were involved in priority setting. Funds for the construction of rural roads were obtained from different sources, such as the state budget, Minimum Needs Programme (now Basic Minimum Services), Central Assistance for Rural Landless Employment Guarantee Programmes, famine and drought relief, and Employment Guarantee Programme. Some of these programmes were later merged into one programme under the name 'Jawahar Rozgar Yojana', abbreviated to JRY.

We will return to this problem later. Meanwhile, the working of the system at the local level is fairly effective. The standard of work achieved in rural and village roads is good. Community participation in the implementation of the construction programme is effective. The villagers normally donate land for the road through their Panchayats. Sometimes, there is a labour contribution; always the villagers are involved in the construction; they keep watch on the material used and the progress of work. Deficiencies are reported to the supervisory authorities and soon become an issue. There is a separate rural roads organization for back-up. Planning of works is on an annual basis and is transparent. The source of construction material at the Naguesh quarry is identified and the rate structure is worked out for each village in the schedule of rates. The real problem now is the identification of additional works given that all villages stand connected. Field responses suggest two reasons for additional demands, economic and religious.

Granite polishing has come up in Sola village on the border of the Ahmedabad Municipal area. Around 134 persons are employed in the cutting and polishing factories. A majority of them commute from two villages which as the crow flies are 3.5 and 5 kilometres away. These are Chainpur and Jagatpur. But during the four-month monsoon period the direct fair weather path is not available and they have to commute more than 12 kilometres using National Highway No. 8 off which Sola is located. On account of the fast expansion of building activity, the work of granite cutting and polishing is expected to expand rapidly.

The village of Ganapatipura is near Dholka a town in rural Ahmedabad. It boasts of a centuries old Ganesh temple, the origins of which are shrouded in antiquity. On Saturdays and "*Chaturthi*", the fourth day of the lunar calendar, up to fifty thousand pilgrims throng to the temple. On special occasions, this crowd can double. The village of Ganapatipura is situated off a district road. All the villages around the temple have a high demand for a direct connection to this religious centre. Two such connections to clusters of villages have already been constructed. One more has just been approved. It is broadly estimated that from diverse sources – people's contributions, employment programmes, contributions from local area development funds of members of the state legislature – from 20 to 25 kilometres of such roads are built annually. But there is no comprehensive plan for demand oriented road construction, which is additional to the All India plan for rural connectivity.

Ahmedabad district is also interesting in that a special bond for financing development works has been issued by the Ahmedabad Municipal Corporation. The Corporation is financing investments, in part, through India's first municipal bond issue. This investment programme is the culmination of a progressive programme of administrative and financial reform within the corporation itself and a partnership developed between USAID, the FIRE Project of USAID and the Ahmedabad Municipal Corporation.

The Ahmedabad project features two important innovations in the urban infrastructure sector. First, it is financed by India's first municipal bond issued without a state guarantee. Ahmedabad's first credit rating, issued in February 1996 was A+, indicating adequate security for investors, for an issue of 1 billion rupees. The Municipal Corporation returned to Credit Rating Information Services of India (CRISIL) the following year, however, with a structured debt obligation, a concept introduced by the FIRE Project, to request an SDO (structured debt obligation) rating. CRISIL issued an AA(so), indicating a high degree of security. With this SDO approach, the corporation was able to draw greater interest in its bonds.

The bond issue was prepared by IL & FS (lease financing agency) whom the corporation had appointed as its investment bankers. Structured basically as a general obligation, it also had a revenue bond flavour due to credit enhancement in the form of a structured repayment mechanism. Specified revenues, in this case receipts from specified octroi collection posts, are deposited in an escrow account and payments to bond holders are supervised by a trustee. The issue was structured as a 75 per cent public placement, with the remainder publicly issued. When the bonds were finally issued in January 1998, they were fully subscribed and regarded as a great success.

The second innovation relates to the management and implementation of the Corporation's infrastructure investment plan. Project management consultants have been enlisted from the private sector to facilitate the process of project design and approval, tendering, construction supervision, quality control, audit and budgetary controls and payments. This structure is based on the approach developed by the City and Industrial Development Corporation of Mumbai to successfully execute a number of large projects simultaneously. The FIRE Project has been worked out to promote the use of project management consultants from the private sector as a means of increasing efficiency, lowering costs and decreasing time and cost overruns. Another innovation has been a 3.5 kilometre market in the prime commercial area of Ahmedabad constructed with private-public partnership. Ahmedabad Municipal Corporation appointed Mr Bimal Patel, an award-winning architect who also runs a non-profit corporate firm of town planning services, to design the most "modern" urban road in the country. A local firm agreed to finance the 35 million rupee project. Advertisement rights of the new road have been given to the firm for seven years to recover its costs.

The Gogalgaon group of five villages belongs to a perpetually drought prone area. Rainfed crops, such as millets, lentils and fodder crops, were grown here. Allied economic activities such as dairying and poultry gave some income stability to the farmers. Dr Vitthalrao Vikhe Patil Cooperative Sugar Factory and the many educational institutions provided employment opportunities. From 1995, the villages were covered under an integrated watershed development project, with the support of PIRENS Loni, an NGO. The project components include farm bunding to prevent rain water and top soil run-off, continuous contour trenches along the hillside and tree plantation in those trenches, grass-growing to encourage stall-feeding of cattle so as to protect the trees, etc. In the first year, the water table improved and there was a shift to growing fruits, vegetables and flowers. About 40 per cent of the land abandoned earlier due to water shortage was brought under cultivation.

The villagers contribute 16 per cent of their daily wage earnings as their project contribution and 100 rupees per year per family as a maintenance fund. The entire project is being implemented with the help and cooperation of a village watershed committee which receives technical and managerial inputs from the project staff and takes the necessary decisions. Once in two months, all the villagers come together to review the progress of the project. They use this platform to discuss conditions, access to health-care facilities and the need for a sports ground. The women of the villages have formed a self-help group. Discussions are ongoing about the economic activities they would like to start.

The first activity undertaken by the villagers, apart from the project work, was filling the craters on the approach road to the project site to facilitate the visits of people to see the progress of the project. They contributed the tractor time, stone and labour. As a goodwill gesture, the NGO met the fuel expenses to run the tractor. They covered the open drains at points where there was maximum seepage. A fairly good road which interconnects three villages and ultimately connects to the Mumbai-Nashik State Highway became the top agenda during village meetings. The road, as proposed by the villagers, would reduce by 10 kilometres the distance to stable markets for the fruits and flowers grown in the villages. It would improve access to the Indian Technology Institute and the arts, science and commerce college in Rahata. Sugar cane growers from Ashwi and nearby areas need to transport cane to the Pravaranagar sugar factory. For them, the distance would now be reduced. Students from Ashwi and nearby villages would have a shorter distance to travel to reach the schools and colleges, including technical, medical and dental colleges in Loni, if the new road was constructed. The speciality hospital in Loni would become more accessible to the villagers. The distance of the villages to Shirdi, a major pilgrim centre, would also be reduced by 10 kilometres. This pilgrimage centre attracts thousands of pilgrims on major Hindu and Muslim festivals.

Roads in a Backward Region

The Constitutional Amendments have set up a legal structure but in many areas the instrumentalities for development planning at the local level have not as yet been activated. The Marathwada area in Maharashtra, which is a backward drought prone area but with a long tradition of local self-government, is an example of this pattern. Recent case studies in different parts of India suggest that own resources as a method of financing development efforts has decreased. This has also meant a decline in the ability of local bodies to raise institutional finance. This, in turn, has meant that the operational flexibility of these institutions has decreased. Their dependence on central resources, particularly through the employment scheme – JRY, has increased. The share of the local bodies' own resources funding recurring expenditure – mainly salaries of staff – has gone up. The overall size of development expenditure has in many cases gone down. In many areas there are too many "starts" of investment projects leading to lack of completion and cost overruns through works in progress. Generally, maintenance of existing assets has been neglected. There has been in many cases no relation between levels of development and the size of the development effort.

POLICY AREAS FOR INSTITUTIONAL DEVELOPMENT

We briefly review the current situation of local institutions for participatory planning in India. The basic constructs of the system have been set up.

Effectiveness depends on three factors. The first is a genuine attempt at giving flexibility to local institutions in implementing major developmental initiatives such as the Basic Minimum Services Programme and the Special Employment Programme (JRY). The second is the need to involve local institutions in the country's economic liberalization and restructuring programmes. The Ahmedabad Municipal Bond issue focuses on this issue, which is basic to the success of participatory planning for infrastructure development in India. Without restructuring, local bodies will not be able to mobilize resources for such development. This restructuring has to be done at three levels – local, state and central. Finally, when the three tier system has been set up, its development will need continued emphasis on transparency, genuine transfer of power and administrative coordination and reform, particularly in planning and implementation.

The first aspect to underline is that while local government, as a part of the process of democratic decentralization, is over four decades old in India, it was the Seventy-third Amendment to the Constitution which was indeed a far-reaching, if not a revolutionary step. It gave constitutional validation and empowerment to the third tier of government. Elected leadership at the local level became mandatory. The provision of positive affirmation mandated through the constitutional law that one third of the elected leadership at the local level had to consist of scheduled castes, scheduled tribes and women, was indeed a major breakthrough. The process of transfer of power has begun and in the long run there will be no going back. The discussion in this paper underlines the fact that there is a need for greater focus on the central resources made available to the Panchayats through JRY. The system of responsibility and resource-raising at the state level has as yet not been set up in an organized way and the question of restructuring and reform needs urgent attention. As regards roads, the problem of rural connectivity is still to be solved in some regions, but in others, the quality of the road link is becoming the issue. Resources for the purpose will involve a restructured system of local finance.

The basic minimum need for roads is an accepted norm in India. The situation at present is that in some large states substantial progress lies ahead. In Arunachal Pradesh, Bihar, Madhya Pradesh, Orissa, Punjab and Rajasthan, less than 40 per cent of villages are connected. The Annual Plan for 1996/97 contained a long-term policy on the provision of seven basic common services throughout the country in a time-bound manner, including rural roads. An allocation of 24.6 billion rupees for basic common services was made under the heading 'Central Assistance to State Plans'.

For this, a group of Chief Ministers working with the author evolved a strategic approach to the State Plan. This indicated development options by the Planning Commission, leaving details to the state authorities, and the stipulation of much stronger evaluation procedures. The Planning Commission indicated the overall ceilings of additional resources to the state and they, in turn, decided on the phasing of priorities in areas such as drinking water, rural literacy, health programmes and rural roads.

The state governments were requested to take similar initiatives towards decentralization at the local level. This approach, which is included in the approach paper to the Ninth Plan, was described by the present author as follows: "Economic growth and employment opportunities in themselves may not be sufficient to improve the living conditions of the poor. They need to be accompanied by measures which enhance the quality of life.

For concretising this approach, a number of steps have already been taken which provide the initial outlines of the larger initiatives that will be contained in the Ninth Five-year Plan. At a conference of Chief Ministers held in July 1996, it was decided to implement a programme for the achievement of total coverage of seven basic minimum services in a time bound manner.

It was agreed to raise the outlays of these programmes by 15 per cent in spite of the stringent fiscal situation. The Ninth Five-year Plan will continue this commitment in real terms through each of the five years. While the objectives of this programme have been decided through the process of mutual consultation, the states have been given full opportunity to decide on the phasing of the target for each specific sector. The achievement of these targets will be jointly monitored by the state and the central governments. This approach is an ideal blend of national commitments with local initiatives. The method of planning for agriculture related infrastructure, irrigation and water planning, and other infrastructure such as power, railways, communications and information technology and science and technology will be a variant of the methods developed for the basic minimum services programme.

Plans will be set up with well defined targets set at the national level in detailed consultation with the states. Policies will be explored in each sector to provide for more investment from the private sector, from cooperatives and voluntary organizations and international private investment. However, the slack in these selected sectors will be taken by public investment. The objectives will be defined nationally but the states and local governments will be given much larger freedom for choice of programmes, phasing of schemes and choice of appropriate instruments of finance".

The system of greater involvement of Panchayats in rural development was institutionalized in 1989/90 with the launching of JRY where there was a substantial flow of funds to the village level, every village having access to JRY funds, though the amount varies. In addition, the village Panchayats prepare an inventory of assets and give details of the projects taken up by them under JRY.

This acts as a social audit. The works to be taken up are decided in the meeting of the Gram Sabha and these are prioritized according to the felt needs of the local people. It is thus possible to take a holistic view of the development needs of the area. This enables the village Panchayat to meet the critical gaps in rural infrastructure and also to generate income via supplementary wage employment to those willing and able to do manual work. It is hoped that with greater people participation in decision-making, formalized through the institutions of Panchayati Raj, there will be greater transparency in the functioning of these institutions eventually leading to fuller accountability. As discussed earlier, apart from the basic minimum services programme, JRY would, if properly implemented, give autonomous resources to people's representative institutions.

This paper suggests that some Panchayats and municipal bodies are able to meet a substantial part of their current expenditure from their own revenue sources. Such bodies should become the norm for others and the state finance commissions could provide the background for a devolution scheme that provided for such a desirable outcome. When this happens, the policies should be developed to permit local bodies to go in for "borrowing" to meet their infrastructure reforms.

At present, successful borrowing schemes, such as the Ahmedabad Municipal Corporation which received a high CRISIL rating, are very few. A financial reform process is required for this purpose so that municipal and panchayat paper is traded in financial markets and India moves towards meeting its large social and economic infrastructure requirements by a strategy in which those communities are helped which decide to organize themselves for this purpose.

Local bodies must have well-defined tax and non-tax bases and autonomy in fixing bases and rates. Their responsibilities in raising revenues have to be clearly defined and these must become a base for public debate and discussion. The costs of irresponsibility must be brought out. The norms of performance can be developed from need, as also from the experience of better performing local bodies. The principles of transfer to the local bodies from the state governments and between the local bodies must be transparent, stable and defined in relation to need and efficiency objectives.

Non-tax user charges are to be emphasized from both the angle of revenue and, more important by from the view of optimal use of scarce resources and environmental and susceptibility criteria. The existing successful experiences need to be documented and replicated. As local bodies move over to more viable financial fundamentals, their capacity to mobilize resources from borrowings will improve. Local bodies have to demonstrate that they can function effectively within a hard budget constraint. They can then use a set of instruments to mobilize borrowings for long-term needs. Tax-free bonds, innovative methods of securing collateral for borrowing for community purposes and design of new financial instruments, in an increasingly developed financial market, should be examined both at the technical and policy level. These last set of institutions have to be set up as a part of national fiscal reform. It must, however, be realized that the process of restructuring will be time-consuming. As such, well-defined short term and medium term targets must be defined and accepted, and progress towards them regularly monitored.

The basic local institutions for participatory planning in India have been set up. However, these institutions ought to have adequate autonomy as units of self-government so as to decide on the local needs and priorities and design and implement the necessary action. The challenge of making this new phase of democratic decentralization successful depends on the commitment of the political leadership, bureaucracy and the people themselves.

The state governments have the responsibility of transferring schemes as provided in the 11th Schedule of the Constitution and also of transferring corresponding funds to the Panchayats. Similarly, the staff available for the implementation of schemes related to these subjects should also be placed under the control of the Panchayats. Sufficiently clear and workable regulations have to be framed to streamline the inter-tier relationships and functions of each tier.

The change which has been initiated by the Seventy-third Constitution Amendment needs to be managed and sustained effectively through innovative strategies. Panchayats have been empowered by this Amendment. Access to and effective control over resources is a critical component of this empowerment. The real issues of such access are knowledge and awareness of the needs of the people, legal rights, and availability and accessibility of social and economic resources. These changes in procedures and organizations have as yet not fully taken place as envisaged in the legislation. It also requires restructuring relationships, including looking at gender issues at both the micro and macro level. A culture of equality has to be evolved not only between men and women representatives but also between officials and non-officials manning the decentralized development structure.

The district has been recognized as the basic unit for decentralized planning functions. Operationalizing the concept of district planning requires functional local institutions. The District Planning Committee is the statutory authority for preparing plans for local development area planning. The involvement of the people in the planning process is necessary to take into account their felt needs, to mobilize local resources, to increase the speed of implementation by securing the people's cooperation, to increase the acceptance of the plan and projects and also to bring about a change in the power structure in people's institutions in favour of the poor. Strong leadership and political will are the necessary conditions for facing the challenge of enabling the local self-government institutions to become effective instruments of social and economic development of rural areas.

5

Rural Education Development

RURAL EDUCATION

Majority of India still lives in villages and so the topic of rural education in India is of utmost importance. A survey named called the Annual Status of Education Report (ASER), shows that even though the number of rural students attending schools is rising, but more than half of the students in fifth grade are unable to read a second grade text book and are not able to solve simple mathematical problems. Not only this, the level of maths and reading is further declining. Though efforts are being made, they are not in the right direction. The reason cited for this problem in surveys is the increasing number of single classroom to educate students from more than one grade. In some states attendance of teachers and students is also declining. These are a few reasons why schools have failed to educate rural India.

Quality and access to education is the major concern in rural schools as there are fewer committed teachers, lack of proper text books and learning material in the schools. Though Government schools exist, but when compared to private schools then quality is a major issue. Majority of people living in villages have understood the importance of education and know that it is the only way to get rid of poverty. But due to lack of money they are not able to send their children to private schools and hence depend upon government schools for education. Above that, in some of the government schools there is only one teacher for the entire school and if they don't show up at work, then it is a holiday. If the quality along with number of teachers and, that too committed teachers can be improved in these schools, then aspiring rural children and India can fulfill their dreams of doing something great.

Some government schools in rural India are overly packed with students, leading to a distorted teacher- student ratio. In one such remote village in Arunachal Pradesh there are more than 300 students in class X which makes nearly 100 students in each classroom. In such a situation it is impossible for teachers to pay full attention towards each and every student, even if they are willing to help.

Every village is not provided with school which means that students have to go to another village to get education. Owing to this parents usually do not send their daughters to school, leading to a failure in achieving rural education in India.

Poverty is another setback. Government schools are not as good and private schools are expensive. This results in a very low number of students actually clearing their secondary education and taking admission in a colleges for further studies. So the drop-out-rate at the secondary level is extremely high in villages. Only parents who can afford college education send their kids to secondary schools. If parents are not able to send their wards for higher education then all their previous efforts get wasted as completing just secondary education means a low paying job and the person is again struck in the same never ending cycle of money, life and poverty.

Most textbooks are in English and since people in rural areas either speak their native language or Hindi, but not English that defeats the purpose. This results in lack of their interest in studies. Though some of the students from villages are really brilliant, as they have a wealth of practical knowledge and know how to survive even in very harsh conditions of life, difficultly in understanding their textbooks, lack of facilities and their poverty are a hurdle in their education.

Quality related issues are far powerful than poverty. Students are not at all encouraged to think but they are asked to memorize pre-defined questions for exams. So for many students clearing examination at the end of the session, passing their exam becomes more important than gaining knowledge. Also as per the new CBSE rule, every student is supposed to be promoted to the next class irrespective of marks in their examination. Hence majority of students do not bother to study, which means a decline in their education level . Neither students nor teachers take any interest in studies which is why the level of education is declining in India despite many efforts.

The foundation to turn India into a strong nation has to be laid down at primary and rural levels and so the quality of education right from the beginning should be excellent. Education and text books should be made interesting. For rural students textbooks related to their culture, their traditions and values should also be there so as to create their interest in studies.The reasons behind so many drop-outs in spite of free education should be found out as this is a hurdle on the road to progress. Improvement in the condition of government schools, education quality, committed teachers and more salaries to these teachers should be part of development.

There is a difference between city and village student not in terms of brain or development but their initial environment, skills, learning ability, availability of infrastructure, and access to different facilities. All of these must be considered while making the curricula which should not be different but how it is going to be taught would make the difference. Encourage the genuine rural students who are interested in education and make them competent. There are many examples of success in rural education in India like the Barefoot college, 8 Day Academy and Gurukul School in Bihar. These are innovative and successful examples of schools running in rural India. It is the time to replicate such efforts as our country and its rural population is very vast which means one of two stories of these kinds won't make any difference. Instead of this large number of such schools are required in rural India. It is also absolutely mandatory to evaluate the success of the schools and students at each and every level. Timely assessment will throw light on present problems and achievements. Let us try to build a solution around these problems which will resolve the overall issues of rural education in India.

REVITALIZING RURAL EDUCATION IN INDIA

In response to a globalizing world that requires increasingly high technological literacy, our own Amrita University, in collaboration with the Indian government, has announced a new pilot program that aims to provide quality computational education to rural children in India. In four village schools across the state of Andhra Pradesh, hundreds of students grade 8 to 10 are participating in workshops that range from robotics to basic life skills.

Dr. Venkat Rangan, Vice Chancellor of Amrita University, explains: "We intend to create enthusiasm among school students about science and technology by making it engaging and fun through digital learning." Since rural locations have limited connectivity, we are working to provide free software and support to these schools as well as 22 others.

"It is very critical to add relevant skills to the schools' curriculum such as computational thinking, robotics skills and maker skills and that is the value that Amrita is bringing to India's Model Schools approach," says Akshay Nagarajan the on-site coordinator of the new program.

"The peer-to-peer interaction going on at these workshops may be the key to successful initiatives in education," says Dr. Sidney Strauss, former Chief Scientist of the Israeli Ministry of Education. Dr. Strauss is working with our volunteers and researchers to design and implement a model that incorporates peer-to-peer interactions into the workshops.

The young participants are gaining unique skills and experiences. For example, our volunteers have set up a cardboard headset that, paired with a smartphone, allows rural students to peek into the world of Virtual Reality.

To balance the technical education, our volunteers are also offering a variety of life skill centered workshops. Amrita University's AMMACHI Labs' Life Enrichment Education program emphasizes the importance of open dialogue about social issues such as hygiene, smoking, alcoholism, and domestic violence.

Our volunteers are hopeful that, as the program expands and develops, more students across rural Andhra Pradesh, especially young female students who are often at a disadvantage educationally, will thrive with the access to high quality education, creative methodologies of learning, and innovative computational training.

THE NATURE OF AN AIM OF EDUCATION

The aim of education is to enable individuals to continue their education — or that the object and reward of learning is continued capacity for growth. Now this idea cannot be applied to all the members of a society except where intercourse of man with man is mutual, and except where there is adequate provision for the reconstruction of social habits and institutions by means of wide stimulation arising from equitably distributed interests.

And this means a democratic society. In our search for aims in education, we are not concerned, therefore, with finding an end outside of the educative process to which education is subordinate. Our whole conception forbids.

We are rather concerned with the contrast which exists when aims belong within the process in which they operate and when they are set up from without. And the latter state of affairs must obtain when social relationships are not equitably balanced. For in that case, some portions of the whole social group will find their aims determined by an external dictation; their aims will not arise from the free growth of their own experience, and their nominal aims will be means to more ulterior ends of others rather than truly their own.

Our first question is to define the nature of an aim so far as it falls within an activity, instead of being furnished from without. We approach the definition by a contrast of mere results with ends. Any exhibition of energy has results. The wind blows about the sands of the desert; the position of the grains is changed.

Here is a result, an effect, but not an end. For there is nothing in the outcome which completes or fulfills what went before it. There is mere spatial redistribution. One state of affairs is just as good as any other. Consequently there is no basis upon which to select an earlier state of affairs as a beginning, a later as an end, and to consider what intervenes as a process of transformation and realization.

Consider for example the activities of bees in contrast with the changes in the sands when the wind blows them about. The results of the bees' actions may be called ends not because they are designed or consciously intended, but because they are true terminations or completions of what has preceded.

When the bees gather pollen and make wax and build cells, each step prepares the way for the next.

When cells are built, the queen lays eggs in them; when eggs are laid, they are sealed and bees brood them and keep them at a temperature required to hatch them. When they are hatched, bees feed the young till they can take care of themselves. Now we are so familiar with such facts, that we are apt to dismiss them on the ground that life and instinct are a kind of miraculous thing anyway.

Thus we fail to note what the essential characteristic of the event is; namely, the significance of the temporal place and order of each element; the way each prior event leads into its successor while the successor takes up what is furnished and utilizes it for some other stage, until we arrive at the end, which, as it were, summarizes and finishes off the process. Since aims relate always to results, the first thing to look to when it is a question of aims, is whether the work assigned possesses intrinsic continuity.

Or is it a mere serial aggregate of acts, first doing one thing and then another? To talk about an educational aim when approximately each act of a pupil is dictated by the teacher, when the only order in the sequence of his acts is that which comes from the assignment of lessons and the giving of directions by another, is to talk nonsense.

It is equally fatal to an aim to permit capricious or discontinuous action in the name of spontaneous self- expression. An aim implies an orderly and ordered activity, one in which the order consists in the progressive completing of a process. Given an activity having a time span and cumulative growth within the time succession, an aim means foresight in advance of the end or possible termination. If bees anticipated the consequences of their activity, if they perceived their end in imaginative foresight, they would have the primary element in an aim. Hence it is nonsense to talk about the aim of education—or any other undertaking—where conditions do not permit of foresight of results, and do not stimulate a person to look ahead to see what the outcome of a given activity is to be. In the next place the aim as a foreseen end gives direction to the activity; it is not an idle view of a mere spectator, but influences the steps taken to reach the end.

The foresight functions in three ways. In the first place, it involves careful observation of the given conditions to see what are the means available for reaching the end, and to discover the hindrances in the way. In the second place, it suggests the proper order or sequence in the use of means. It facilitates an economical selection and arrangement. In the third place, it makes choice of alternatives possible. If we can predict the outcome of acting this way or that, we can then compare the value of the two courses of action; we can pass judgment upon their relative desirability.

If we know that stagnant water breeds mosquitoes and that they are likely to carry disease, we can, disliking that anticipated result, take steps to avert it. Since we do not anticipate results as mere intellectual onlookers, but as persons concerned in the outcome, we are partakers in the process which produces the result. We intervene to bring about this result or that. Of course these three points are closely connected with one another. We can definitely foresee results only as we make careful scrutiny of present conditions, and the importance of the outcome supplies the motive for observations.

The more adequate our observations, the more varied is the scene of conditions and obstructions that presents itself, and the more numerous are the alternatives between which choice may be made. In turn, the more numerous the recognized possibilities of the situation, or alternatives of action, the more meaning does the chosen activity possess, and the more flexibly controllable is it.

Where only a single outcome has been thought of, the mind has nothing else to think of; the meaning attaching to the act is limited. One only steams ahead toward the mark. Sometimes such a narrow course may be effective. But if unexpected difficulties offer themselves, one has not as many resources at command as if he had chosen the same line of action after a broader survey of the possibilities of the field. He cannot make needed readjustments readily.

The net conclusion is that acting with an aim is all one with acting intelligently. To foresee a terminus of an act is to have a basis upon which to observe, to select, and to order objects and our own capacities. To do these things means to have a mind — for mind is precisely intentional purposeful activity controlled by perception of facts and their relationships to one another.

To have a mind to do a thing is to foresee a future possibility; it is to have a plan for its accomplishment; it is to note the means which make the plan capable of execution and the obstructions in the way, — or, if it is really a mind to do the thing and not a vague aspiration — it is to have a plan which takes account of resources and difficulties.

Mind is capacity to refer present conditions to future results, and future consequences to present conditions. And these traits are just what is meant by having an aim or a purpose.

A man is stupid or blind or unintelligent — lacking in mind — just in the degree in which in any activity he does not know what he is about, namely, the probable consequences of his acts.

A man is imperfectly intelligent when he contents himself with looser guesses about the outcome than is needful, just taking a chance with his luck, or when he forms plans apart from study of the actual conditions, including his own capacities. Such relative absence of mind means to make our feelings the measure of what is to happen. To be intelligent we must "stop, look, listen" in making the plan of an activity.

To identify acting with an aim and intelligent activity is enough to show its value — its function in experience. We are only too given to making an entity out of the abstract noun "consciousness." We forget that it comes from the adjective "conscious." To be conscious is to be aware of what we are about; conscious signifies the deliberate, observant, planning traits of activity. Consciousness is nothing which we have which gazes idly on the scene around one or which has impressions made upon it by physical things; it is a name for the purposeful quality of an activity, for the fact that it is directed by an aim. Put the other way about, to have an aim is to act with meaning, not like an automatic machine; it is to mean to do something and to perceive the meaning of things in the light of that intent.

RE-IMAGINING RURAL EDUCATION IN DIGITAL INDIA

The gloomy state of educational structure in India has been a distressing issue for past several decades. Though many schemes have been launched to improve the scenario, the condition of rural education in India continues to be miserable. Issues like - lack of proper infrastructure, the dearth of professionally trained teachers willing to work in remote locations, gender differentiation and difficulty in attracting students to school - continue to hinder the rural education.

As a result, there is a huge gap between the already-attained and yet-to-be-attained learning goals. E-learning can be the key to unlocking the challenges and help India bridge this gap, specifically in rural areas.

E-learning transform the face of rural education in India

Elearning is simply a term used to encapsulate 'the use of technology for *effective* learning', and what could be more *effective* than deploying this tool for spreading the light of education in the darkest corners of the nation!

E-learning's invasion in rural education has emerged as both - a ray of hope as well as a challenge. In a developing country like India, e-learning undoubtedly offers great opportunities to empower the rural areas.

With Information and Communication Technology (ICT) crafting e-learning's path, it can incredibly transform the face of rural learning. Here are some facts that back the bright future of eLearning in India:

Educational Technology and Rural India

Educational technology gives access to a large amount of digital learning material. A good Learning Management System (LMS) shapes the information and adds learnability quotient to it. The world of digital learning is vast, but with the right 'knowledge management tools' it is possible to design an excellent 'student-centered course'.

Though the process requires intense planning and precise implementation, education technology is certainly capable of overcoming the bottlenecks hampering the progress in rural learning. Thus, the overall all socio-economic impact will be remarkable:

Quality Teaching

Good teachers can be made available to students in remote areas through video conferencing or virtual guidance.

Infrastructure

With online courses, students in rural India can learn even in the absence of a physical school.

Reusability

By utilizing off-the-shelf content, a lot of time and effort can be saved. The content (video lectures, printable notes, app, etc.) once created can be reused multiple times to teach in different regions.

Student retention

Talk-n-chalk classrooms if replaced by PCs or portable devices connected to the internet may play a major role in attracting as well as retaining more students.

Freedom to learn

With digital learning, students can choose from a diverse variety of courses and learn what they are interested in.

Flexibility

As online resources can be accessed anytime, anywhere – even the students who are unable to attend regular classes can study. This will help students who have to work or are differently abled.

Analytics

The course outcomes can be assessed quickly with minimal efforts and used for formulating future policies.

The Hurdles

In an ideal world, e-learning can be designed to reach the rural masses and perfectly overcome the glitches in traditional learning. But the ground reality is much different!

While e-learning brings the promise of improved teaching methodologies, it also comes with a number of concerns related to its execution.

Though eLearning holds the potential to transform the ideal scenario into a reality, these factors are constraining it:

Language

A major percentage of Indian population is incapable of reading or writing English, and translating all the content into Hindi or Regional languages can be a daunting task.

Connectivity

Seamless internet connectivity is still not available in many parts of the country. With the number of internet users expected to double by 2018, meeting the bandwidth requirements will be a bigger challenge.

Digital Literacy

Despite a sudden increase in digital literacy, many are still not conversant in understanding digital terminologies and devices.

Awareness

Spreading awareness about eLearning among the uneducated rural population is another 'difficult to tackle' problem.

Tracking and Security

Verifying the examiner and examinee authenticity, as well as tracking the course completion need implementation of extra mechanisms besides the learning process.

mLearning - Reaching the rural India

The steep rise in the number of mobile users and internet connectivity reaching rural India - the stage is set for mLearning. According to a recent report by Internet and Mobile Association of India and IMRB-International, "Rural India recorded 87 million mobile Internet users, doubling the number in 2014.

The rural has a large potential for mobile internet and the data consumption is poised to grow leaps and bounds."

Smartphones - Changing the ground realities

Smartphones are penetrating deeper in rural areas as the prices are dropping constantly. As the smartphone owners are increasing, the horizon of mLearning is broadening. While setting up educational institutes may take years, mLearning can reach the current rural generation with immediate effect.

Going offline - Resolving connectivity issues

Besides having the usual eLearning benefits, mLearning brings in solutions which allow accessing the content offline in a secure manner. This resolves the issue of not having seamless internet connectivity. The data is sent back to the server when connected to the internet, thus smoothly tracking the course progress and collecting feedback.

The Initiatives - A step forward

Call it National Literacy Mission, Right to Education, or Sarva Shiksha Abhiyan (SSA) – different names were given at different times however the goal was one EDUCATION FOR ALL! As a result of tremendous efforts from the Government and several NGOs, the status of rural education has improved drastically.

Digital India

The Government of India (GoI) under PM Modi's directive has been taking huge steps to support eLearning. According to a recent report by the UK-India Business Council, "India has become the largest market for e-learning after the US, and the sector is expected to receive a boost from the government's Rs. 1.13 trillion Digital India initiative."

CSR (Corporate social responsibility)

Organizations like HP, Cognizant, Azim Premji Foundation, HCL, Cisco, etc. have taken up initiatives to educate the rural India via digital methods. As eLearning itself is still in its nascent stage not much can be predicted about its future role in rural education, but there's no denying the fact that 'eLearning is the ideal solution for all the difficulties faced by the education system in rural India.'

Several Non-Profit organizations are educating teachers as well as students in remote locations in India using CloodOn platform.CloodOn sees a clear shift in usage from desktops to mobile devices. There is also an increase in usage of rich content like videos facilitated by CloodOn and partner mobile apps.

THE CHALLENGES OF EDUCATION IN RURAL INDIA

When we talk about education in India, we can't just talk about how education is in urban cities of India, without going deep into rural education that constitutes almost 90% of the schools being located in rural areas. Recent studies have shown how the face of education in rural parts of the country have developed to a great extent, but some remote areas still do need a serious checkup with children failing to receive basic quality education.

We cannot but agree that, it is education that leads to the road to betterment of a community and the nation at large. And when we think about bring in a reformation in education, we have to point out what all prevents the education system in India to develop. Let's start it with rural education. The most common problems that hinders the growth of education in rural India can be pointed out as

- Lack of proper transportation. Most villages have poor connectivity from one place to another and that is often one of the main reason why, despite efforts by local governing bodies to build schools, often go in vain. Children, most of the time have to walk miles to reach these government funded schools and this often demotivate them to attend school on a regular basis.
- People belonging to remote rural areas have meager incomes, which at times is too less to sustain a family of maybe four or five. Most likely, children from these families won't be sent to schools, instead would be asked to assist the earning member of the family to add up some

extra income. On the other hand, teachers in rural educational centers in villages are paid poorly, often leading to lack of attention by teachers, ultimately forcing the students to suffer.

- Lack of proper infrastructure at these rural schools is also a big concern. Most of the schools don't have proper classrooms, teaching equipment, playgrounds and even basic facilities like clean toilets. Thus, the poor condition of schools are big reasons to drive away students.

These are some of the few prominent crisis that's holding back rural education to match up with the education system in urban educational centers. Education imparted in rural centers lack in quality and it's high time, proper attention is paid to these centers and create a platform where students from rural areas can get proper education, the right encouragement to pursue further studies and jobs.

Proper availability of basic facilities like clean toilets, drinking water, adequate classroom facilities, timely motivational programs for teachers etc should be there. The right reformation can definitely bring about a positive change towards the development of rural education in India.

DEVVEOPMENT OF TECHNOLOGY IN RURAL EDUCATION OF INDIA

Mobile phones, internet, tablets, iPads, their applications, social media even traveling, cooking, communication etc are part of our lives from the start till the end of the day. Technology is touching every aspect of society and changing it dramatically. But there is one very important and indispensable part of the society that has also been tapped by new innovations and discoveries and that is education. Like all other areas, in this case also urban areas are influenced to a greater extent than rural one. So much more could have been done to bring the revolution in learning process in rural areas of India.

In India illiteracy is one of the biggest problems. Lack of easy access, lack of teachers, lack of interest, poverty, gender differentiation, lack of infrastructure, common curricula are few of the reasons which are holding back the progress in rural education. But with the use of technology mass education can be given and situation can be changed.

To reach rural areas, first of all study material can be distributed to the students then online interaction and online videos can be made with teachers. Online teaching creates extended classroom communities for discussions, virtual classrooms and for interaction.

There is another option in which classroom courses can be recorded in a real time and used for teaching the students who cannot attend these classes.

This creates an expanded access to education. Rural education needs e-learning technologies. Apart from this audio conferencing and video conferencing should be made part of the education system in rural India. Teachers at the schools are not well equipped with the gadgets. So teachers should be given printers, laptops, for giving notes and notices to the students. By using technology the problem of unqualified teachers can also be solved.

Initiatives

Public private partnerships are playing wonderful role in making rural India tech savvy and in providing education. Successful initiatives have been taken and some are even in pipeline. NGOs are playing an important role in technology assisted rural education.

A not-for-profit organization, Azim Premji Foundation, run by Wipro group has been working towards this issue since 2001. It is helping 2 million children in 16,000 schools from 14 state governments. This foundation works by assisting computer-aided learning. Computers are considered as an asset. These are major draw to bring students to the school. For children computer is a very exciting machine.

Schools in the states like Andhra Pradesh, Assam, Chhattisgarh, Himachal Pradesh, Karnataka, Madhya Pradesh, Meghalaya, Tamil Nadu, Tripura and West Bengal have received computer-assisted education through collective efforts by NIIT and government. This has given positive results. Dropout rates have reduced to a great extent through computer-aided education.

Edusat – Provided video education in rural India. Edusat was launched by ISRO.

VidyaGyan – Works with an aim to uplift the exceptional rural students from economically disadvantage background. They are given world-class education free of cost at VidyaGyan which is a residential institute. It is an initiative of the Shiv Nadar Foundation founded by Shiv Nadar. He is the founder of HCL. VidyaGyan has taken 200 students who scored the highest in the UP state board examinations. From sixth grade onwards these students will be taught at VidyaGyan.

Samudaya: Under this programme 2000 children were taught English, social science, science and mathematics through technology. Using same method 250 teachers were also trained and the outcome was really positive. It tremendously reduced the school dropout rates and absenteeism and at the same time improved the level of confidence. Government has to create more e learning centers to educate students living in rural areas. Required e-learning courses can then be defined. Such e-learning systems will surely improve the society both economically as well as socially. They will develop and come at par with urban students.

Hurdles

Rural India is deficit in technology. Along with this rural India severely suffers from the shortage of water, infrastructure, power, health facilities etc. But for the success of every technology based education, infrastructure has to be strong. Discontinuation of electricity is the major problem out of all. There is an option for the same. Solar energy is available in abundance that can be used in many cases.

Access to education is second major issue. Students have to cover miles for education. For this more schools are required to be built. Schools should open on the basis of population, and distance.

Advantage

Technology attracts the rural children. Their wish to attend school increases with the word computer. Even parents show more willingness in this case. Computer education builds up the required confidence and narrows down the gap between urban and rural education. So rural India needs great investment in this chapter.

PROBLEMS ASSOCIATED WITH EDUCATION IN RURAL AREAS IN INDIA

The concept and phenomenon of education based on school-going is of modern origin in India. Education in the past was restricted to upper castes and the content taught was also ascriptive. However, today, to lead a comfortable life in this fast-changing world, education is seen as the most influential agent of modernization.

The educational attainments in terms of enrolment and retention in rural India gen-erally correspond to the hierarchical order. While the upper castes have traditionally enjoyed and are enjoying these advantages, the Scheduled Caste and other backward castes children have lagged behind in primary schooling. Studies have revealed that chil-dren of backward castes are withdrawn from school at an early age, by about 8 or 9 years.

An important reason for withdrawal of children from school is the cost and work needs of poor households. Income and caste are typically correlated with lower castes having lower incomes and higher castes having better endowments in terms of land, income and other resources. Thus, one fact is certain that there is a clear divide in the villages, along caste lines, regarding access to schools. The very poor children are enrolled in the municipal school because it provides a number of incentives such as lower expenditure on books, uniforms, fees, etc. The well-off children go to the private school, where English and computers are given more importance.

The tendency in favour of private schools in rural areas is influenced by people's perception of private schools, as a means of imparting quality education in English medium. The poor rural girls, if not all, con-stitute a major junk of disadvantaged groups that are excluded from the schooling pro-cess, especially because they enter late into school and drop out earlier.

Parental illiteracy is another cause for lack of interest to become literates. Many rural children enrolled are thus first generation learners, who come from illiterate families thus, they have to single handedly grapple with school life, mastering language and cog-nitive skills without parental help and guidance.

Most of these illiterate parents do what-ever is possible to educate their children because education for them acts as a vehicle of social mobility. Moreover, education and the subsequent attainment of town jobs is often looked upon by many of these rural families, especially families belonging to lower castes, as a means to break out of their position in caste hierarchy.

The religious beliefs and practices of a community can also largely impact the overall attitudinal and behavioural profile of an individual or group. In the Indian context, reli-gion has a sway over people's minds and exerts a great influence over their behaviour. The motivation and attitudes of the people towards education are also moulded, to a large extent, by their religious beliefs. The literacy rate for Muslims is notably lower com-pared to Hindus but not better than Christians and Sikhs. Poverty among Muslims, who also happen to be one of the most economically back-ward groups, is the actual reason for their preference for madrasas (Islamic schools), because they are absolutely free and more flexible as compared to formal government schools. This seems to be the only option for poor Muslims, who often cannot afford to pay for the education.

Defects of Present System

According to Amartya Sen, 'Primary education in India suffers not only from inadequate allocation of resource, but often enough also from terrible management and organiza-tion. To him, management and organization of schools is still in a terrible State in India.

That means, there are three major defects in the present educational sys-tem. The first is the physical environment in which the student is taught, the second is the curriculum or the content, which he/she is taught, and the third is the teaching method or the teacher, who is teaching.

Physical Environment

Today's society clings to schools to such an extent that a co-dependent relationship is created between the broader and friendly notion of education and the manipulative real-ity of school. Education should not be limited to the sphere of the school.

It should have to encompass nearly every aspect of life. Schools should act as locations where the ideas of education are planted in the students and education has to become the foundation for how the students look at the world around them and how they interpret these things. Instead the present situation is that, the seeds of education are planted into the children in the schools but it does not go much further than the school system.

The public in general and rural people in particular, often think of schools as a place for teachers to instruct children on the 'three Rs'—reading, writing and arithmetic. Schools are not con-sidered as places, where the students are taught many life skills that will help them suc-ceed in their future endeavors.

Access to school is no more a problem in most parts of India. Ninety eight percent of population has access to school within a walking distance of 1 km. The core problem is the unpreparedness of the school system for mass education. Classrooms in most pri-mary schools in rural areas are typically uninviting, with leaking roofs, uneven floors and scraggly mats to sit on.

Added to that, most of the schools do not have electricity, drinking water or toilets. In some schools, students of different ages are made to sit in one room. These students squat in passive postures, even regimented columns, with often the 'brightest' and the socially advantaged sitting in front. At a given time, a typical school could have at most two teachers trying to 'police' children of all five primary classes.

The best teaching that these teachers may undertake is to make the students copy or recite from the textbook. Sounds emanating from the school are normally distin-guishable from afar in the form of a ritual cacophonous chorus of children chanting their lesson, often shouting their guts out in a cathartic release.

Surprisingly, no normal sounds— of joyous laughter, creative play of words, singing or recitation of poems, ani-mated participation, excited discovery, or even the irrepressible curious questioning the characteristic of every child of that age are found in schools.

The major drawback in these schools is that in the mechanical race to achieve 'schooling for all' the government seems to have completely missed out on what constitutes 'learning for all'. Here, greater emphasis is placed on establishing schools but not on what goes on inside a school.

The result is high enrolment figure and equally high dropout rate. The students enrolled are compelled to attend school regularly and take all the exams, and the result is a sizeable number of students fail and are compelled to repeat classes.

These students ultimately give-up the hope, resulting in high resource wastage of the government, while at the same time inculcating a sense of despair among the students, thus, reducing the poten-tial of their human development.

The quality of education is the main issue. For a long time, the educationists had thought that the high dropout rate is because of parental poverty and disinterestedness rather than concentrating on the failure of the school system.

A paper by Jandhyala B. G. Tilak, titled 'Determinants of Household Expenditure on Education in Rural India, which is based on the results of NCAER's 1994 Human Development in India (HDI) survey, tries to clarify some of such myths that are associated with the education in rural India. The study mentions that the real household expenditures on education in India is not 'virtually non-existent', but is considerably higher.

Some of the observations made by this study are as follows:

1. First, there is a complete absence of 'free education' in India, regardless of a household's socio-economic background, spending on education is very sub-stantial even at the primary school level.
2. Second, 'indirect' costs, such as books, uniforms and examination fees, are very high, even in government-run schools, including at the primary level. According to National Sample Survey Organization (NSSO), in 1995-96, the average expenditure per student pursuing primary education in rural India in a govern-ment school was Rs.219, and for students going to local body schools, private-aided schools and private-unaided schools, it was Rs 223, Rs 622 and Rs 911, respectively.
3. Third, given the absence of a well-developed credit market for education, expenditure on education is highly (and positively) correlated with income.
4. Fourth, willingness to pay and 'compulsion to pay' (i.e., the need to compensate for a shortage of government spending on education) are two important factors.

5. Fifth, government spending and household spending on education are not sub-stitutes but complementary. An increase in government spending is associated with an increase in household spending (due to an enthusiasm effect, resulting from improvements in school facilities, number of teachers, etc.).

   Conversely, a reduction in government expenditure leads to a decline in household spending on education. (Equivalently, the elasticity of household expenditure to govern-ment expenditure is found to be almost unitary, and positive.)
6. Finally, the provision of schooling in rural habitations, or the provision of such school incentives as mid-day meals, uniforms, textbooks, etc., are both associ-ated with the increased household demand for education.

Defects in Curriculum

The second reason attributed to high dropout rate is 'most out-of school children are unable to study because they have to work'. The recent study by PROBE has refuted this reason as myth. It says that only a small minority of children are full-time wage labour-ers, while the majority of those, who work do so as family labourers at home or in the fields.

According to this study, it is not because the children have to work that they leave school, but because they leave school they work. Then why do children leave schools?

The present education system is teacher-centric. From times immemorial, education has been expensive, as it is related to gaining and transmitting knowledge in India. 'Information,' the foundation upon which knowledge rests was and still is in limited supply.

A teacher, together with a united set of books, is the knowledge base, which anchors the education process. The teacher is the active agent, communicating information to the students, who are the passive receptors of information.

Learning by rote is the method most favoured because the information transmitted is largely disjointed and the student is not really quite sure what the motivation behind knowing all those disparate facts is.

Curriculum-makers in India feel that children need to know a lot more to catch-up' with the others living in advanced countries.

So, they try to include as much 'informa-tion and 'knowledge' as possible in the curriculum. While it is true that this century has seen an explosion of technologies that help to store large information, the capacity to understand these facts and concepts does not grow equally fast among the children. Therefore, a crucial aim of children's education should be to promote concept formation and enhance their capacity for theory-building.

All children are natural theory builders, and from very early in life, they begin to construct their own explanations for the world they observe. The educational system presents a contrasting experience to these chil-dren.

These outdated school systems do not allow for a child's mind and personality to develop. Moreover, the knowledge imparted is not continuous and are disjointed frag-ments of information that are arranged in the form of different pieces in the syllabus.

The curriculum-framers, while arranging such information, ignore the fact that the natural learning process in children is far from linear, and that they process information about the world in a much more holistic and integrated manner. The content taught, therefore, cannot be determined by what has to be 'covered' in higher classes, but by the children's ability to comprehend the concept at a given age.

In the Indian system of education, what is taught is crucially linked with how it is taught and, more importantly, with how it is assessed. The examination system here is so distorted that it actually discourages good classroom practices by forcing children to answer contrived meaningless questions, suppressing their own curiosity and expres-sion.

It emphasizes on written questions based on trivial recall, and discounts all activity- based learning. Another major drawback is that the curriculum prescribed in the textbooks to a great extent are found to be 'irrelevant' to the closer lives of rural people.

The curriculum-makers, who come predominantly from urban middle-class background believe that the rural children need to be taught how to conduct their lives 'properly', and that only 'positive' situations from their lives must be depicted.

Thus, either highly pre-scriptive and moralistic lessons (about hygiene, cleanliness, hard work, etc.) or rather simplistic generalizations about the perceived 'needs' of the rural poor or over-idealized situations such as truly democratic panchayats, benevolent employers, well-equipped and functioning village hospitals, effective government schemes, etc., are described in the prescribed textbooks.

They absolutely ignore the fact that the rural children, unlike those from protected urban homes, are much more conscious of the conflicts and com-plexities of life, which form a part of their reality.

They know very well that these lessons are contrived and untrue, but have no chance to critically question the contents that they must passively parrot.

The village child is also far more knowledgeable about the natural world, and does not need to look at 'pictures' to count the legs of a spider, or to identify the eggs of a frog or the leaves of a neem tree.

Similarly, the teachers and other school authorities also neglect the fact that many rural children, especially the tribal children, are aware of rich bio-diversity around them. They become conscious only when some foreign companies pirate this information and patent it. Thus, in the name of relevance, most of the content taught in the schools is totally irrelevant to the prevailing situations of the rural children.

In recent years, there have been attempts to change the elementary school curricu-lum to make it more child-centred, joyful and activity-based. However, in most cases, there have been mere cosmetic changes accompanied by much 'song and dance', with no radical restructuring in the content and design. The education systems in this region are highly monopolistic and rigid, and are controlled by bureaucratic departments that are resistant to change.

Moreover, the people, who design school curricula have outdated notions about what constitutes 'learning', are burdened with the perceived demands of higher education, and are far removed from a typical average child of the country.

They are also far too inflexible to learn from the experience of village teachers and field-based voluntary groups working in close contact with children. Some flexibility may be per-mitted in the curriculum for 'those children' in non-formal schools, but never in the highly guarded formal system for 'our kids'.

Most Important Resource—The Teacher

The protagonist of the educational system and the most important resource for quality education, the teacher, in reality has the feeblest voice in the matters of concern. The rural primary school teacher occupies the lowest position in the hierarchy. Apart from teaching, he/she is expected to bear the burden of many other assignments such as col-lecting census, propagating family planning programmes, conducting poverty surveys, etc.

The teacher in a village acts as the sole multi-purpose village functionary, and is expected to perform whatever function the government finds necessary at any time. This problem becomes most acute in the case of village schools, having a single or at most two teachers. For days together, the school may remain closed because the teacher has been summoned on 'duty', further discouraging children, who in the absence of support at home, need much more attention and extra time.

Another major problem that has come up in recent days is that, due to the ceiling on recruitments, there is insufficient number of teachers in many of these schools. A con-siderable proportion of available teachers do not have the requisite qualifications. Women teachers constitute only 31 percent of the total number of teachers in rural areas, which is supposed to be one of the causes for low literacy rate.

Lack of motivation on the part of teachers, who generally are academically low- qualified and have chosen this profession as a last resort, is a serious problem. Added to this, the teachers are burdened with unmanageable classes, irrelevant curricula, dismal working conditions, and lack of recognition of their efforts. Moreover, these teachers are lowly paid. The result is that the teachers tend to give up.

6

Training for Social Work and Rural Development

TRAINING FOR SOCIAL WORK

The contribution which social services make will depend to a considerable extent upon personnel and leadership. A general understanding of the philosophy and history of social work, the structure and functions of society, the nature and extent of social problems, the methods and techniques of social work, and of the details of the programmes and how best their results may be assessed, will help improve the quality and efficacy of all services organized by State and private agencies. The training of social workers should of course include knowledge of conditions prevailing in fields in which they are to work, and social workers must possess the spirit of service and the character and energy to execute programmes despite handicaps and limitations and with such resources-as may be readily available.

There are several schools of social work in India and the setting up of some other institutions on similar lines is being contemplated in some of the States. There are important problems involved in these institutions which require specially qualified and experienced personnel, careful selection of candidates for training, special training for fields in which there is scope for employment, and adequate opportunities for field-work experience. Trained social workers are needed in large numbers for rural areas. It should be possible for the existing schools of social work to draw students from rural areas and to arrange for their training in the field in selected centres organized by rural welfare agencies. Universities and colleges in or near rural areas could also develop training programmes for rural development. Agricultural colleges could introduce intensive social welfare courses and field-work 'programmes as part of their curricula. Similar institutions with greater emphasis on social anthropology could be created in tribal' areas.

It is not possible for many voluntary organizations in the country to employ highly trained personnel for their ordinary programmes and activities. It is, therefore, necessary to arrange for training at the community level for field workers, instructors and supervisors. The existing schools of social work, specialized social service agencies, social welfare agencies functioning at the national and State level should provide opportunities for such training. Arrangements for ' in-service' training should also be made by the larger voluntary organizations which have worked in the field of social welfare for many years. Further, arrangements have to te made for the training of voluntary workers who will be needed in large numbers during the coming years. It is especially desirable that voluntary administrative and field personnel should be given some elementary training in social work.

The emergence of State social services and of large central organizations to deal with important social problems and the lack of opportunities for higher training in the social sciences within the country indicate the need in selected cases for training and study abroad in specialized fields. It is necessary that persons who go abroad for training should first have sufficient knowledge and experience of Indian conditions and problems.

Suggestions for Social Development

There are several important issues which require detailed deliberation in social development.

The Issues are:-

1. Considering the fast growth of social development sector in 21st century there is need for Government of India to take a proactive role for multisectoral coordination and convergence of various role players involved in social development for which several Ministries of Govt. of India are required to be sensitized.
2. Need for greater transparency and Accountability in the sector. The world's leading human rights, environmental and social development international organizations such as Action Aid International, AMNESTY International, and Green Peace International, Oxfam International, Save the Children International and World YWCA have today publicly endorsed the first global accountability charter for the non-profit sector to act as responsible players for social development.

   International NGOs play an increasingly influential role. Global public opinion surveys show higher trust in NGOs than in government and business. In addition to an internal desire to be transparent and accountable, the accountability charter also seeks to demonstrate that NGOs deeply value public trust

In an unprecedented step, international civil society organizations have come together to demonstrate their commitment to transparency and accountability. This initiative builds on the individual, national and sectoral initiatives taken by international NGOs to set standards of accountability and codes of conduct.

3. Need for greater respect, appreciation and support to the social workers. The reason for this they are the key actors for the overall development. Every part of social development sector depends upon their ability and responsibility. Efficient and effective social workers will be made by the concerned NGOs. NPOs, NGOs and other developmental organizations and foundations are needed to respect and appreciate the social workers.

   The lack of social workers and the continuous high turnover of social workers weaken the quality and availability of the service given to clients especially for the poor. The lack of competent personnel may lead to the loss of the basic social rights intended in the constitution. Reasons for the lack of social workers are low pay, lack of leadership and heavy workload.

   In order to improve the availability of social workers and their quality work it is suggested that social workers should have the opportunity to participate in continuing education and supervision of work.

4. Need for further promotive policies of activities self groups, micro credit & market avenues for the productivity self help groups. NGOs can promote the policies for self help groups and micro finance. There is strong relation between the self help groups and micro credit system at present scenario.

   In India self help groups are extensively working as primary tools towards poverty alleviation and empowerment. National and state government initiatives, as well as NGOs efforts, have used SHGs to implement poverty alleviation programmes in Andhra Pradesh since 1979. Micro credit is emerged as strong weapon to eradicate poverty through self sustainable with the help of NGOs, micro finance institutions and banks. At present scenario of social development with relation to

5. Need for clear distinction and clarity of Micro Credit & Micro Finance. Mainly SHGs are the prime clients for the Micro and Micro Finance. Banks or financial institutions need to clarify the difference between these two things, because banks are main sources for the micro credit or finance.

   Micro credit and micro fiancé, often used synonymously, is very popular terms in recent developmental activities. This is creating huge confusion and misunderstanding in developmental activities. Professor Muhammad Yunus mentioned this problem with some sarcasm in his address to the International Seminar on Attacking Poverty with Micro Credit, held in Dhaka on 8 and 9 January 2003: The word micro credit did not exist before the seventies. Now it has become a catchword developmental practitioner. In the process, the word now means everything to everybody.

Micro credit caters commercial needs of poor for enabling them to raise their income levels and improve standard of living. Micro credit means more emphasis on loans while micro fiancé also includes support services where you open up channels for thrift, market assistance, technical assistance, capacity building, insurance, social and cultural programmes. So where there is micro finance is credit plus, there only micro credit is credit.

Micro credit financial requirements are generally not meant for economic development activities, but for consumptive needs like it education of a child, medicinal requirements etc. Here quantum are quite low, needs are very emergent, and there is hardly any difference between the consumptive purpose and productive purpose.

6. Need for coordinating Agencies & mechanism both at Govt. of India and at various State Govt. levels. Need for coordinating Agencies and mechanism both at Government of India and at various State Government level. Coordinating agencies and mechanism helps in brining the all NGOs together and facilitate for better work in social development with the help of government machineries.

   More over, these instruments helps to form networks at all level including international, national, state and local level. This body can also coordinate NGO movements in each country.

7. Need for enabling policies both by centre and states. The social problems of contemporary India are the result of a complex nexus between the factors of exclusion and inclusion rooted in history, values, and cultural ethos. Many of these problems could not addressed by the development strategy launched since independence. Recent policies of globalization have further undermined the role larger societal norms as well as the state apparatus that could counter exclusionary forces. The agenda of social development has remained unfinished, keeping social tensions simmering.

   During the 7th five-year plan, polices were helped to achieved the targeted social development goals, in terms of establishment of social infrastructure, especially in rural areas. The 8th five year plan identified "human development" as its main focus, with health and population control listed as two of six priority objectives. It was emphasized that health facilities must reach the entire population by the end of the 8th plan. The plans also identified people initiative and participation as a key element. With the enactment of the 73rd Constitutional Amendment Act (1992), Panchayati Raj Institutions (PRIs) were revitalized and a process of democratic decentralization ushered in, with similar provisions made for urban local bodies, municipalities and nagar palikas.

Today, however, in the policy debate, ongoing orthodox economic liberalism is giving way to concerns regarding social consequences of globalization, as it affects the poorest and the marginalized sections of the population. Thus, a number of highly important and far-reaching social policy measures have been brought on to the development agenda, in the form of the right to information act, rural employment guarantee act, the rural health mission among others.

8. Need for capacity building of Govt. Officials, NGOs and gross root level activists and stake holders of social development. Several international and national conferences identify an effective leadership role of the NGOs as social development channels.

   Capacity building can be defined as "development, fostering and support of infrastructure, resources and relationships for NGOs and related systems and services, at Member States, organizational, inter-organizational, and regional and systems levels, contributing to the peaceful, socially distributed and sustainable development of our societies."

   Capacity-building programmes broaden and strengthen the professional expertise and accelerate progress in the organization activities weather it government or non governmental organization. Capacity building's main goals are to increase the individual capacity of present and future developmental professionals and leaders and to support the development of institutions and programmes all over the world especially in the social development sector.

9. Need for process documentation, Action research, Monitoring & Evaluation and social audit.

   The word "documentation" includes both records and documents. Records are recorded information, regardless of the medium or characteristics, made or received by an organization that is useful in the operation of the organization. Documents explain what an organization plans to do and how it will be accomplished as well as instruct employees how to perform tasks. In this regard professional organizations need to maintain quality and effective documentation along with the action research, monitoring and evaluation and social audit.

Action research is very important component for the effective functioning of the organization. Action research finds out the achievement and failures of the organizations activities. With the help of action research findings, NGOs and other developmental organizations can reform or modify their on going developmental projects, if there are any insufficient methods or policies, social audit is also one of the important elements for the social development. NGOs are the prime players to enhance the awareness about the social audit within the NGOs and other governmental organizations.

10. Need for further promotion of participatory approaches of the development. Participatory approach of the development facilitates the local communities play a central role in the planning, implementation and funding of activities within participatory developmental programmes. The exact composition of any given programme should be determined in conjunction with them. It is important to ensure that programme activities:
    1. do not provoke conflict between resource users (where conflict is unavoidable, conflict resolution mechanisms should be specified early on);
    2. do not further isolate marginal households (that may not be able to participate in activities which demand a labour or financial contribution);
    3. do not undermine viable indigenous soil and water conservation techniques;
    4. are informed by an understanding of existing management practices (e.g. they do not immediately promote group activity if there is no history of communal working);
    5. are feasible given current capacity within the community and external organizations; and
    6. take into account underlying climatic, hydrological, soil and land use characteristics.

    Participatory approaches are more important to succeed the developmental programmes. Participatory approach also enables the social capital for social development.
11. Need for sharing of knowledge, innovative approaches, and people centred approaches. Here, sharing of knowledge is very important component in every aspect of development field. This concept of sharing

of knowledge leads to innovative methods in development sector with development organizations. People centred approach, at present, is very much needed in the field of social development. These above components considered as critical elements for speedy social development. NGOs and other developmental organizations need to develop an innovative approach for social development with the help of sharing of knowledge concept. To get access of sharing of knowledge and to develop innovative approach, NGOs and other developmental organizations need to get extensive trainings on various issues such as Organizational Development, Capacity Building of Organizations and other related to social development. Moreover, NGOs are considered by international developmental organizations and banking as the prime actors for social development.

12. Need for savings, economy of resources and no cost, low cost and cost effective approaches. There is interrelation between the above concepts such as savings, economy of resources and no cost, low cost and cost effective approaches in social development. Developing the society with all minimum needs required the tremendous savings in all sections of the society. Especially in rural areas, women self help groups savings are reached at maximum level. This saving came from only marginalized sections of the society. In this scenario, NGOs and other developmental organization's efforts utilized cent percent with no cost and low cost effective approaches. Thus, for overall social development need the above concepts. Here, NGOs and other voluntary or developmental organizations need to enhance the awareness on above issues for better social development.
13. Need for exchange visits, study tours, & National and International exposure. Exchange visits and study tours at national and international exposure will broaden and strengthen the organizational capacity in various aspects such as management skills, developmental methods and interaction with developmental professionals in the field of national and international and so on. Moreover, tours and visits also give knowledge in the different areas of social development from different nations especially from developed countries. The developed nation's technological knowledge will help in faster the social development in developing countries.
14. Need for planning by NGOs working in the sector. As earlier stated by international developmental organizations, NGOs are the critical players in social development. Because, NGOs works in the every aspect of society and they works from grass roots level to international level. So, NGOs can draw the plans for social development effectively.

15. Need for introduction of managerial inputs to the NGOs working in the sector. For the effective programme implementation and policy-making, NGOs need to acquire the knowledge of effective managerial inputs for the NGOs in the different areas of social development. In the present scenario of social development, every programme of poverty alleviation in rural as well as urban areas and infrastructure development such as roads, drinking water, schools and health centres etc., are implemented and monitored by NGOs. Thus, for effective implementation or benefits reaching to the poor are depends up on NGOs effectiveness.
16. Need for sustainable development approaches. Need for collaboration among peoples institutions, NGOs and Government Institutions. The international community has recognized the vital importance of cooperation between government agencies and non-governmental organizations (NGOs) in addressing the social and human development related issues. The importance of government-NGO cooperation was stressed in the recent global and regional conferences concerning the social development. To respond to these mandates, international developmental organizations formulated the process of government-NGO collaboration for social development especially in the area of poverty alleviation programmes. Moreover, World Bank also is stressed on the Government and NGO collaboration for rapid social development in recent international conferences.
17. Need for Multisectoral coordination and convergence of various role players in the sector. Multi sector coordination and convergence is the effective method for rapid social development. Governments across the world are grappling with appropriate policies to optimize the benefits associated with convergence through multi sectoral coordination. Convergence has emerged as a global phenomenon as a result of digitization which has allowed traditionally distinct services to be offered across interchangeable platforms. These technological trends have been accelerated by the liberalization of markets allowing for the social development.
18. Need for integrated Micro Planning at village level and Holistic development. Integrated Micro planning is the most important concept for the developmental and governmental organizations. In this aspect, every resource of the village utilized for the overall development of the village through the micro planning. In this process, natural resources and human resources utilized for overall development with the help of NGOs and government organizations.

19. Need for creation of the Model Villages. Model villages especially created for the utilization of information and communication technology. In the previous years, ICT was utilized for the development of corporate industries and other large size multi national corporations. Now, in the globalized era, ICT is reaching to the every corner of the globe including villages. If ICT need to utilize properly in village development, villages need to be developed as model villages so as to accept the access of developed technologies. NGOs and other developmental organizations need to be developed the facilities in rural areas with coordination with the government organizations.
20. Need for non exploitation, lack of corruption, and lack of hypocritic approaches by a few black sheep in social development. As late Prime Minister of India, Rajeev Gandhi noticed that exploitation and corruption are the major hurdles for the path of India's development. This is indicating that the high intention and impact in the social development. NGOs and other developmental organizations and charities and every social group are need to work for the non exploitation, lack of corruption, and lack of hypocritic approaches for the better society and social development.

The inter-relationship between the various activities has to be emphasized and the necessary coordination assured both in the Central Government; and in the States. One aspect of this coordination would be to secure that legislation relating to social problems follows broadly similar principles. In cases where grants-in-aid are given by a State authority to a private agency, it is desirable to lay down general directions for improving the content of the programmes and their administration. A measure of supervision and inspection should also be provided in order to maintain standards of efficiency. A major responsibility for organizing activities in different fields of social welfare, like the welfare of women and children, social education, community organization, etc., falls naturally on private voluntary agencies. These private agencies have for long been working in their own humble way and without adequate State aid for the achievement of their objectives with their own leadership, organization and resources.

Any plan for the social and economic regeneration of the country should take into account the service rendered by these private agencies and the State should give them the maximum cooperation in strengthening their efforts. Public cooperation, through these voluntary social service organizations, is capable of yielding valuable results in canalizing private effort for the promotion of social welfare. One of the most important tasks of the State is to conduct a survey of the nature, quality and extent of service rendered by voluntary agencies in different parts of the country, to assess the extent of financial and other aid that they are in need of in order to develop their programmes of work, and to coordinate their activities.

A sum of Rs. 4 crores has been provided as grants-in-aid to voluntary social service organizations for strengthening, improving and extending the existing activities in the field of social welfare and for developing new programmes and carrying out pilot projects. It is envisaged that this fund of Rs. 4 crores should be administered by a board to be set up by the Central Government to which a great deal of administrative authority will be devolved. The board should be predominantly composed of non-officials who have actual experience of field work in promoting voluntary welfare activities.

PRINCIPLES OF SOCIAL WORK

The social work profession is guided by a distinct set of abstract values and a Code of Ethics. These values are transformed into accepted practice principles for the purpose of informing our intervention with clients. What follows is a listing of nine Social Work Principles and brief description of each.

Acceptance-Acceptance is a fundamental social work principle that implies a sincere understanding of clients. Acceptance is conveyed in the professional relationship through the expression of genuine concern, receptive listening, intentional responses that acknowledge the other person's point of view, and the creation of a climate of mutual respect.

Affirming Individuality-To affirm a client's individuality is to recognize and appreciate the unique qualities of that client. It means to "begin where the client is." Clients expect personalized understanding and undivided attention from professionals. Individualization requires freedom from bias and prejudice, an avoidance of labelling and stereotyping, a recognition and appreciation of diversity, and knowledge of human behaviour. Purposeful Expression of Feelings-Clients need to have opportunities to express their feelings freely to the social worker. As social workers, we must go beyond "just the facts" to uncover the underlying feelings.

Non-judgmentalism-Communicating non-judgmentalism is essential to developing a relationship with any client. It does not imply that social workers do not make decisions; rather it implies a non blaming attitude and behaviour. Social workers judge others as neither good or bad nor as worthy or unworthy.

Objectivity-Closely related to non-judgmentalism, objectivity is the principle of examining situations without bias. To be objective in their observations and understanding, social workers must avoid injecting personal feelings and prejudices in relationships with clients. Controlled Emotional

Involvement-There are three components to a controlled emotional response to a client's situation: sensitivity to expressed or unexpressed feelings, and understanding based on knowledge of human behaviour, and a response guided by knowledge and purpose. The social worker should not respond in a way that conveys coldness or lack of interest while at the same time cannot over identify with the client.

Self-Determination-The principle of self-determination is based on the recognition of the right and need of clients to freedom in making their own choices and decisions. Social workers have a responsibility to create a working relationship in which choice can be exercised. Access to Resources-Social workers are implored to assure that everyone has the necessary resources, services, and opportunities; to pay attention to expanding choices, and opportunities for the oppressed and disadvantages; and to advocate for policy and legislative changes that improve social conditions and promote social justice.

Confidentiality-Confidentiality or the right to privacy implies that clients must give expressed consent before information such as their identity, the content of discussions held with them, one's professional opinion about them, or their record is disclosed.

DEFINITION OF SOCIAL WORK

The social work profession promotes social change, problem solving in human relationships and the empowerment and liberation of people to enhance well-being. Utilising theories of human behaviour and social systems, social work intervenes at the points where people interact with their environments. Principles of human rights and social justice are fundamental to social work.

Social work in its various forms addresses the multiple, complex transactions between people and their environments. Its mission is to enable all people to develop their full potential, enrich their lives, and prevent dysfunction. Professional social work is focused on problem solving and change. As such, social workers are change agents in society and in the lives of the individuals, families and communities they serve. Social work is an interrelated system of values, theory and practice.

Values

Social work grew out of humanitarian and democratic ideals, and its values are based on respect for the equality, worth, and dignity of all people. Since its beginnings over a century ago, social work practice has focused on meeting human needs and developing human potential. Human rights and social justice serve as the motivation and justification for social work action. In solidarity with those who are disadvantaged, the profession strives to alleviate poverty and to liberate vulnerable and oppressed people in order to promote social inclusion. Social work values are embodied in the profession's national and international codes of ethics.

Theory

Social work bases its methodology on a systematic body of evidence-based knowledge derived from research and practice evaluation, including local and indigenous knowledge specific to its context. It recognises the complexity of interactions between human beings and their environment, and the capacity of people both to be affected by and to alter the multiple influences upon them including bio-psychosocial factors. The social work profession draws on theories of human development and behaviour and social systems to analyse complex situations and to facilitate individual, organisational, social and cultural changes.

Practice

Social work addresses the barriers, inequities and injustices that exist in society. It responds to crises and emergencies as well as to everyday personal and social problems.

Social work utilises a variety of skills, techniques, and activities consistent with its holistic focus on persons and their environments. Social work interventions range from primarily person-focused psychosocial processes to involvement in social policy, planning and development.

These include counselling, clinical social work, group work, social pedagogical work, and family treatment and therapy as well as efforts to help people obtain services and resources in the community.

Interventions also include agency administration, community organisation and engaging in social and political action to impact social policy and economic development. The holistic focus of social work is universal, but the priorities of social work practice will vary from country to country and from time to time depending on cultural, historical, and socio-economic conditions.

METHODS OF SOCIAL WORK

Social work as a profession is a product of this century. Although its roots are well established in history from the time when people 1st began to take responsibility for their neighbours through activities which were called charity, poor relief, philanthropy and social reform.

Social work is to fight against Five Evils as:

1. Physical want
2. Disease
3. Ignorance
4. Squalor
5. Idleness.

Objective:

- To remove social injustice
- To relieve social injustice
- To reduce redress
- To prevent suffering
- To assist the weaker sections
- To rehabilitate the distress class people.

Social Case Work

Social case work is a method which helps by counselling the individual client to effect better social relationships & a social adjustment that makes it possible him to lead a satisfying & useful life. Gordon Hamilton points out that, "The objective of case work is to administer practical services & offer counselling in such a way as to arouse & conserve psychological energies of the client activity to involve him in the use of the service towards the solution of her/his dilemma."

Social Group Work

Social group work is an activity which helps to participate in the activities of a group for their intellectual, emotional & physical growth and for the attainment of desirable goals of the groups.

Group work as such as a method by which the group worker enables various types of groups to function in such a manner that both group interaction & programme activities contribute to the growth of the individual & the programme activities contribute to the growth of the individual & the achievement of desirable social goals.

Community Organization

Community organization is the process of planning & developing social services in order to meet the health & welfare needs of a community or larger unit. Mildred Barry says," Community organization in social work is the process of creating & maintaining a progressively more effective adjustment between community resources & community welfare needs."

Social Action

It is an organized group process solving general social problems & furthering social welfare objectives by legislative, social, health or economic progress. The term social action refers to organized & legally permitted activities designed to mobiles public opinion, legislation & public administration in favour of objectives believed to be socially desirable.

Social Welfare Research

Social welfare research systematic critically investigation of questions in the social welfare field with the purpose of yielding answares to problems of social work & of extending generally social work concept. The methods appliesd in social work research have been to a largwe extent derived grom those used in sociology & social psychology as well as in history and Anthropology.

Social Welfare Administration

Social welfare administration process is to organize & to direct a social agency. The administrative aspects of social work have to do with the organization & management of social agencies public & private, including in those terms general administrative relationships among units of the same organization, personal problems, questions of finance & so on.

DEFINING SOCIAL WORK

Any attempt to explore the role and function of the social worker must of necessity seek to answer or at the very least address one basic question - what is social work? Our concern here is not to indulge in semantics though there is a serious issue about the language within which social work is framed and to which we will return later. Rather, our purpose is to identify the main definitions of social work as a social movement. The logic for this is very simple. Without some agreed notion of what it is that constitutes social work, it is well nigh impossible to determine what the role of the social worker might be; what skills and expertise social workers should have; what training and education is appropriate for social workers; and what it is that distinguishes social work from other professions.

Differences in what is taken to be social work can also account in part for the gap between what social workers say they want to achieve and accomplish and what they are able to achieve within the constraints of the institutional settings within which they operate. It may also partly account for the gap in expectations of those who seek to enter the profession of social work and what they experience when they are in a work situations. It may also account for the fact that materials and documentation relating to the training courses offered by training institutions rarely (though with some exceptions) offer intending students any definition of what social work is.

Cree states that 'It is almost impossible to find a simple definition of social work with which everyone is likely to agree'. Thompson suggests 'Social work is what social workers do'. For Cree, though the Thomson position is not seen by her to be very helpful, goes on to state that: We should not expect to find unanimity in books about social work, or even in accounts of social workers. Social work is always subject to competing claims of definition and practice, and cannot be separated from the society in which it is located. Rather social work has to be seen as a collection of competing and contradictory discourses that come together at a particular moment in time to frame the task of social work.

Characteristics of Social Work

We live in a challenging, fascinating world. We have learned how to travel to the moon and project a satellite around Mars, yet we often have difficulty reaching someone in the same room. We live in a fast-moving, push-button space age; yet, we find that people are still the center of the world and its values. The rocks, the stars, and the moon are important; but humans are more important. As people live together, problems of relationship and interaction emerge. Personal problems, family problems, and community problems appear on the horizon of everyday living. Drug abuse, crime, delinquency, mental illness, suicide, school dropouts, and numerous other social problems abound at every turn. Every day new scientific and technological discoveries enlarge our knowledge and skills. We not only travel in space but also have one or more cars in the garage for personal travel.

We carry tiny radios and TV sets that bring instant news 24 hours a day, shrinking the world to a small community. Fifteen-second ads channeled into our homes are common. Wonder drugs are used by millions everywhere searching for peace of mind as well as for treatment. Automatic cameras, VCR's, and computer chips are altering our recreational and educational endeavors at every turn. In 1986 President Reagan and Soviet leader Gorbachev, for the first time in history, sent New Year's messages from America to Russia and from Russia to America to millions of people with opposite political systems and beliefs, raising hopes for increased understanding and peace in years ahead. Medical triumphs surprise the imagination with use of lasers and miniature instruments under high magnification to restore eyesight, use of an artificial arm that gives amputees lifelike dexterity, and test-tube babies.

Although the Shuttle Challenger exploded in February 1986 with the loss of seven astronauts, new space efforts are moving ahead. Serious consideration is even being given to the United States and Russia sponsoring a joint human expedition to Mars for exploration in the 1990s.

On the negative side, all kinds of difficult situations exist. Terrorism is rampant as never before; no place on the globe is any longer free from attack. Fifty-five thousand nuclear weapons are arched ready to be fired at an instant's notice. Single mothers are raising children in many women-dominated homes with no fathers around. Use of drugs is on the increase around the world, bringing bizarre, uncertain behaviour into the lives of people, their families and their communities. Violence on TV increased in America more than 100 per cent between 1980 and 1986. The average child views TV about four hours a day. By the age of 16 he or she has watched more than 200,000 acts of violence—50,000 of which are murders. In this high-tech uncertain world personal, family, and community problems exist as never before. Emotional difficulties are rampant. Teen-age suicides are on the increase. There seems to be an overall ebb in meaningful human relationships—people and helping each other. When serious difficulties in human relationships arise, the question immediately becomes, "Can we get help?"

The answer is usually, "Yes." Many professions, one of which is social work, are ready and eager to help. What is social work? What is social welfare? What part do they play in this satellite age? Social work and social welfare are based on three premises: (1) that the person is important, (2) that he or she has personal, family, and community problems resulting from interaction with others, and (3) that something can be done to alleviate these problems and enrich the individual's living. An introductory glimpse of one particular case illustrates all three premises.

> *Joe had been in a state hospital for nineteen years, confined to one of the backwards. Those who had known him reported he was institutionalised because of severe emotional pressures due to a "shattered marriage." The attendants indicated he had talked to no one for several years and that he apparently lived in a world of his own.*
>
> *Then came a change. As a part of experimentation in treatment, a social worker was assigned to meet with Joe for thirty minutes daily with the goal of bringing him back to reality. The worker explained to Joe that she was going to see him regularly and was genuinely interested in him. For the first two weeks, he uttered not a word. Then one day he surprised the worker by saying, "I like you."*
>
> *The worker continued to spend about thirty minutes daily with Joe. She was sympathetic, interested in him, warm, friendly, and accepting. He responded*

steadily to her interest and efforts. Within six months, he was discharged from the hospital and had become established in his community. A follow-up several years showed he had made a successful return to society and was an accepted, productive citizen.

When asked what had happened by a close friend, Joe eagerly replied, but with seriousness, "The worker was just like a magnet—drawing me out of my shell. I couldn't resist her."

Social Work

The next question that follows is, "What is social work?" An answer appeared in an editorial in the *New York Times*:

> *A new profession has been growing to maturity under our noses. It is social work. Once thought of as basket-on-the-arm assistance to the poor, it now is a discipline, scientific in method and artful in manner, that takes remedial action on problems in several ares of society. It ministers to families in economic or emotional difficulty. It helps communities to bring their welfare and related services into good balance. It works in medical, group, and school situations. It seeks to correct the causes underlying delinquency and adult criminality.*

A much-used definition of social work appears in the Curriculum Study, sponsored by the Council on Social Work Education: "Social work seeks to enhance the social functioning of individuals, singly and in groups, by activities focused upon their social relationships which constitute the interaction between man and his environment.

These activities can be grouped into three functions: restoration of impaired capacity, provision of individual and social resources, and prevention of social dysfunction. The underlying assumptions of social work are:

1. Social work like all other professions has problem-solving functions.
2. Social work practice is an art with a scientific and value foundation.
3. Social work as a profession came into being and continues to develop because it meets human needs and aspirations recognised by society....
4. Social work practice takes its values from those held by the society of which it is a part. However, its values are not necessarily or altogether those universally or predominantly held or practiced in society.
5. The scientific base of social work consists of three types of knowledge: (a) tested knowledge, (b) hypothetical knowledge that requires transformation into tested knowledge, and (c) assumptive knowledge (or "practice wisdom") that requires transformation into hypothetical and thence into tested knowledge....

6. The knowledge needed for social work practice is determined by its goals and functions and the problems it seeks to solve.
7. The internalisation of professional knowledge and values is a vital characteristic of the professional social worker since he is himself the instrument of professional help.
8. Professional skill is expressed in the activities of the social worker....

The Curriculum Study describes some of the essential values of social work that are important in defining this profession:

1. Each person has the right to self-fulfillment, deriving from his inherent capacity and thrust toward that goal.
2. Each person has the obligation, as a member of society, to seek ways of self-fulfillment that contribute to the common good.
3. Society has the obligation to facilitate the self-fulfillment of the individual and the right to enrichment through the contributions of its individual members.
4. Each person requires for the harmonious development of his powers socially provided and socially safeguarded opportunities for satisfying his basic needs in the physical, psychological, economic, cultural, aesthetic, and spiritual realms.
5. As society becomes more complex and interdependent, increasingly specialised social organization is required to facilitate the individual's efforts at self-realisation....
6. To permit both self-realisation and contribution to society by the individual, social organization must make available socially sanctioned and socially provided devices for needs satisfaction as wide in range, variety, and quality as the general welfare allows.

According to the Curriculum Study, "These values constitute a minimum commitment for the social worker. They imply definition of human freedom as conditioned by the exigencies of modern living." The goal of social work, according to the Curriculum Study, is the enhancement of social functioning wherever the need for such enhancement is either socially or individually perceived. "The patterns, directions, quality and outcomes of man's social relationships (social interaction) in the performance of his various roles (social functioning) become the professional concern of social work.

A problem presented in the area of social interaction, whether raised as a problem by the individual or by a group in the community, calls for the professional services of the social worker." The basic functions of social work—restoration, provision of resources, and prevention—are intertwined and interdependent. Restoration of impaired social functioning may be subdivided into curative and rehabilitative aspects.

Its curative aspects are to eliminate factors that have caused breakdown of functioning, and its rehabilitative aspects, to reorganize and rebuild interactional patterns. Illustrations of restoration would include assistance in obtaining a hearing aid for a partially deaf child or helping a rejected lonely child to be placed in a foster home.

The rehabilitative aspect might be helping the one child to psychologically accept and live with the hearing aid and supporting the other child as he or she adjusts to the new foster home. Provision of resources, social and individual, for more effective social functioning may be subdivided into developmental and educational. The developmental aspects are designed to further the effectiveness of existing social resources or to bring to full flower personal capacity for more effective social interaction. An example would be the services of a Family Service Society that help Mr. and Mrs. X, through individual and conjoint interviews, to understand each other better and to open the channels of communication between them for the first time in their twelve years of marriage. The educational spectrum is designed to acquaint the public with specific conditions and needs for new or changing social resources. Again, this could be illustrated by public talks given by staff members of a Family Service Society, in which counseling services are described as a resource in alleviating marriage and family problems. The third function, prevention of social dysfunction, involves early discovery, control, and elimination of conditions and situations that potentially could hamper effective social functioning.

The two main divisions are prevention of problems in the area of interaction between individuals and groups, and second, the prevention of social ills. Premarital counseling would be an example of an attempt to prevent individual and social problems in relation to social functioning. Hopefully, through this process couples will be able to anticipate possible difficulties in marital interaction and, through adequate consideration and understanding, avoid the problems that might ensue otherwise. Prevention of social ills ordinarily falls within the area of community organization. An example of this function is the Community Services Council approach to the reduction of juvenile delinquency through the utilisation of all community organizations and economic resources (for example, to build a new Youth Center and provide it with a professionally trained staff to work with boys and girls who are near-delinquent or who live in "delinquency areas"). An overall conclusion in the Curriculum Study states, "that the focus on social relationships, however, is suggested as the *distinguishing characteristic* of the social work profession." Bartlett claims that social work is "a configuration of elements none of which is unique but which, in combination, represent a contribution quite distinct from that rendered by any other profession."

A Current Definition

Social work may be defined as an art, a science, a profession that helps people to solve personal, groups (especially family), and community problems and to attain satisfying personal, group, and community relationships through social work practice, including casework, group work, community organization, administration, and research. Social work practice today is often generic, involving all three of the traditional methods. The major focus is on reducing problems in human relationships and on enriching living through improved human interaction. Certainly the main focus of the social worker is upon helping people to improve their social functioning, their ability to interact and relate to others. On the other hand, there are many in the related helping professions who also assist with interactional problems. In addition, social workers sometimes help individuals to solve individual and personal problems. The social workers ordinarily works with clients on a conscious level, helping them to face realities and to solve problems without delving into the realm of the unconscious. Social work is an art; it requires great skills to understand people and to help them to help themselves.

It is beginning science because of its problem-solving method and its attempt to be objective in ascertaining facts and in developing principles and operational concepts. It is a profession because it encompasses the attributes of a profession.

Distinguishing Characteristics of Social Work

Comprehension of social work may be enhanced by consideration of some of its distinguishing characteristics that follow:

1. Focus is on the wholeness and totality of the person—encompassing the person, environmental factors, and behaviour. Social work stresses the total person in the total environment.
2. Emphasis is on the importance of the family in molding and influencing behaviour. Social workers attempt to understand the principles underlying family interaction and to work with the family as the basic unit for improving social functioning, recognising that most social problems inhere in inadequate or imbalanced family relationships. The family is often regarded as the *case* in social work. Although the modern family is changing and many new forms of marital and family living have appeared, the family is still the basic institution in society and as such is a focal factor in social work.
3. Utilisation of community resources in helping people to solve problems is very important. Social workers have a comprehensive knowledge of community resources and are able to tap them to meet the needs

of their clients. They help to get "Mr. or Mrs. Jones to the clinic," to the agency that can help the most in the solution of his or her particular problems. Making referrals is a major service of social workers. They also make significant contributions to the larger community, utilising their skills in planning and organizing, helping governmental and private organizations and agencies to be more effective.

4. Use of the supervisory process provides for guidance and direction of inexperienced workers and for continuing growth of the experienced. Both in academic study and in practice, social work provides supervision by qualified, professional personnel to help the worker to continue to grow professionally and to acquire increased understanding and skill. The supervisor is available regularly to help the worker do a better job and increase his or her understandings and skills in working with people. This process is particularly important because social workers themselves are the tools in helping troubled persons, and they need to grow professionally, keeping abreast of new knowledge and skills. Patterns of supervision have been changing so they are less rigid today than previously. The general trend is toward more self-direction and less formalised supervisor-supervisee relationships. Participatory supervision, self-supervision, and peer supervision are being used extensively.
5. Social work has a unique educational programme involving classwork and practical field work experience, which go hand in hand. To obtain a Master of Social Work (MSW) degree requires two years of graduate training—or its equivalent—in one of the 88 graduate schools in the United States. This programme includes both academic classes and live field experiences in working with clients, which gives an integrated combination of theory and practice. Provision has been made for undergraduate programmes in social work and/or social welfare to be accredited by the Council on Social Work Education. Thus, graduates in these programmes with a BSW (Bachelor of Social Work) degree are recognised as beginning social workers, qualified to start in professional practice.
6. Traditional social work emphasises three basic processes: casework, group work, and community organization. Casework involves a close, face-to-face relationship—mainly on an individual-to-individual basis—in working with people and their problems. Group work utilises the group as the tool to bring about desired changes in social functioning with troubled persons. Community organization is the intergroup approach toward facing and solving social pathologies. It aims to increase understanding of community needs and helps to provide for

them. Social workers often play an advocate role to strengthen and improve community resources and bring desired social changes. Some social workers claim there is only one social work process—that of problem-solving related to social relationships. They state that the method is basically the same whether working with individuals, groups, or communities.

7. Social work has distinctive professional bodies: The National Association of Social Workers (NASW) and the Council on Social Work Education (CSWE). NASW was established in 1955, after careful study, and brought together several smaller professional social work groups in a unified, dynamic organization. Its membership has increased rapidly and in 1988 it enrolled more than 100,000 social workers. This organization is doing much to raise the standards of social work practice, to recruit qualified persons for professional training, and to interpret practice and values of social work to the public. CSWE, originated in 1952, is helping to improve training facilities, standards, and programmes as well as helping with recruitment, public relations, and strengthening social work practice.
8. The *relationship* is the key in the social work process. Everything that is a part of the interview is important, but to the social worker, the feeling tones between the worker and client are particularly important. The social worker attempts to make it possible for the client to face and solve his or her problems by sharing knowledge and helping with understanding and acceptance in an emotionally supportive relationship.
9. Social work has an orientation in psychiatric concepts and places considerable stress upon understanding people. The social worker is particularly interested in how clients feel about themselves and their relationships with others. The social workers possesses considerable understanding of the basic knowledge and concepts of psychiatry and dynamic psychology that assist him or her in dealing with human behaviour.
10. The *social* in social work emphasised stress on social interaction and resultant social functioning and malfunctioning. Significant principles from sociology and social psychology, as well as from group dynamics, are woven into the artful fabric of social work, and are utilised in understanding relationships of people and in helping them to resolve their conflicts.
11. Social work recognises that social problems and human behaviour inhere to a considerable degree in the social institutions of mankind.

To understand these problems and behaviour, it is necessary to understand the institutions of humans. Social problems may be reduced by working with individual personalities or by changing social institutions. For example, a particular boy may be helped to turn from delinquent behaviour through individual therapy; on the other hand, social work recognises that perhaps thousands of delinquent acts may be prevented through sensible changes in political or economic institutions.

12. Most social workers are employed in agency settings. Although the number of social workers in private practice is increasing, most operate within the framework and policies of agencies; this gives them structural backing and support that strengthens their services in many ways. Supervision, consultation, and collaboration, inherent in agency settings, provide many positive resources for the workers.
13. The basic aim of social work is to help the client to help him or herself or to help a community to help itself. Contrary to what many people believe, the social worker does not listen to a client and then prescribe a "social-psychological pill"—even though many clients ask for this. This social worker endeavors to help a person to improve his or her understanding of him- or herself and relationships with others and to tap his or her own and community resources in solving personal problems. The social worker operates under the premise that most people have the ego strength to solve their own problems when they really bring them out into the open and understand what they are. Whereas several professions are primarily concerned with pathological problems, social work endeavors to stress and utilise strengths—both individual and community—to effect desired changes. Social work also makes a major thrust to understand community resources and to help solve community problems, thereby bringing desired changes.
14. Since most social workers are employed in agencies and are on fixed salaries, fees are utilised for the welfare of the agency rather than for increased incomes for the workers.
15. Traditionally social workers have provided services and therapy for individuals and families. Within the past two decades, prevention has received considerable emphasis, and recently a focus on enrichment in living, for all people, has surfaced.
16. A social worker is particularly effective in developing and using the team approach and in bringing about coordination of services and activities. Many professional workers regard the social worker as the catalyst who has the ability and responsibility to help the professional

team work together and function in optimal fashion. The social worker often acts as coordinator and integrator for the team effort.

SOCIAL WORK AS A CONTESTED CONCEPT

From this perspective, social work is what social work is seen to legitimately be.

This of course presents no statement as to the actual nature and function of social work. It does though emphasise the importance that what social work is, is subject to a process of what Askeland and Payne call 'validation'. That is, a process of legitimation validates what is seen to be the nature of social work from a variety of conflicting definitions and assumptions. Social work is a truly contested concept and at any point in time there are competing definitions. What is clear is that what social work is taken to be has as much been about whose definition is seen as legitimate rather than which definition.

Social work has always been subject to competing claims of definition and practice, as social workers, politicians, service users and policy makers have struggled to lay claims on what social work is, and what it might be.

And, To understand social work, therefore, we must understand how knowledge is validated within the profession.

What this inevitably implies is that there is no universal body of knowledge for social workers. What is seen to be valid knowledge or indeed the function of social work is defined by many others outwith the profession including academics, educators, professionals, administrators, politicians, users, carers and the media. There can be no doubt that within these different constituencies, there are very different views and assumptions about social work and its function, fuelled by vested interests and media representation, especially of problematic cases and scenarios.

This of course makes it very difficult to identify what are the appropriate skills and expertise needed for social work. It also accounts for the fact that social work as a movement has since its earliest days been associated with continual change and critical reflection on what it is; how best and where best it can be exercised. It also makes it very difficult to establish a clear professional identity for social work when the concept itself is subject to the views and assumptions of competing constituencies. Similarly, accepted or validated notions of social work which are embedded in the organisational structures of public social services may be entirely unacceptable to those with a more radical bent. For some, the resolution of this near chaos in competing statements about the nature and function of social work resides in the importance of social workers themselves determining what their specific professional identity is in order to ward off conflicting, indeed detrimental, notions about social work and its role.

NATURE OF THE WORK

Social work is a profession for those with a strong desire to help improve people's lives. *Social workers* assist people by helping them cope with and solve issues in their everyday lives, such as family and personal problems and dealing with relationships. Some social workers help clients who face a disability, life-threatening disease, social problem, such as inadequate housing, unemployment, or substance abuse.

Social workers also assist families that have serious domestic conflicts, sometimes involving child or spousal abuse. Additionally, they may conduct research, advocate for improved services, or become involved in planning or policy development. Many social workers specialize in serving a particular population or working in a specific setting. In all settings, these workers may also be called *licensed clinical social workers,* if they hold the appropriate State mandated license.

Child, family, and school social workers provide social services and assistance to improve the social and psychological functioning of children and their families. Workers in this field assess their client's needs and offer assistance to improve their situation. This often includes coordinating available services to assist a child or family. They may assist single parents in finding day care, arrange adoptions, or help find foster homes for neglected, abandoned, or abused children. These workers may specialize in working with a particular problem, population or setting, such as child protective services, adoption, homelessness, domestic violence, or foster care.

In schools, social workers often serve as the link between students' families and the school, working with parents, guardians, teachers, and other school officials to ensure that students reach their academic and personal potential. They also assist students in dealing with stress or emotional problems. Many school social workers work directly with children with disabilities and their families. In addition, they address problems such as misbehavior, truancy, teenage pregnancy, and drug and alcohol problems and advise teachers on how to cope with difficult students. School social workers may teach workshops to entire classes on topics like conflict resolution.

Child, family, and school social workers may be known as child welfare social workers, family services social workers, or child protective services social workers. These workers often work for individual and family services agencies, schools, or State or local governments.

Medical and public health social workers provide psychosocial support to individuals, families, or vulnerable populations so they can cope with chronic, acute, or terminal illnesses, such as Alzheimer's disease, cancer, or AIDS. They also advise family caregivers, counsel patients, and help plan for patients' needs after discharge from hospitals. They may arrange for at-home services, such as meals-on-wheels or home care. Some work on interdisciplinary teams that evaluate certain kinds of patients, such as geriatric or organ transplant patients.

Some specialize in services for senior citizens and their families. These social workers may run support groups for the adult children of aging parents. Also, they may assess, coordinate, and monitor services such as housing, transportation, and long-term care. These workers may be known as gerontological social workers.

Medical and public health social workers may work for hospitals, nursing and personal care facilities, individual and family services agencies, or local governments.

Mental health and substance abuse social workers assess and treat individuals with mental illness or substance abuse problems. Such services include individual and group therapy, outreach, crisis intervention, social rehabilitation, and teaching skills needed for everyday living. They also may help plan for supportive services to ease clients' return to the community when leaving in-patient facilities.

They may provide services to assist family members of those who suffer from addiction or other mental health issues. These workers may work in outpatient facilities, where clients come in for treatment and then leave, or in inpatient programs, where patients reside at the facility.

Some mental health and substance social workers may work in employee-assistance programs. In this setting, they may help people cope with job-related pressures or with personal problems that affect the quality of their work. Other social workers work in private practice, where they are employed directly by the client. These social workers may be known as *clinical social workers*, *occupational social workers*, or *substance abuse social workers.*

Other types of social workers include *social work administrators, researchers, planners* and *policymakers*, who develop and implement programs to address issues such as child abuse, homelessness, substance abuse, poverty, and violence. These workers research and analyse policies, programs, and regulations. They identify social problems and suggest legislative and other solutions. They may help raise funds or write grants to support these programs.

SOCIAL WORKER FIELDS

Social workers take on the awesome responsibility of helping those who struggle in the world to stay afloat. As a social worker, you will be a vital bridge for someone in need to accessing services and support for the provision of basic needs. This article is a look at the various areas of social work and the educational requirements for going into a career as a social worker.

What is a Social Worker?

Broadly speaking, social work is for people who want to help people, by giving them access to the resources they need to improve their lives and build brighter futures.

Social workers try to help people make the most of their surroundings, their relationships, and their struggles with money or family. Many social workers deal with people who face life-threatening situations, such as criminal activity or substance abuse. Other issues that social workers try to address are inadequate housing, unemployment, illness, disability, or difficulties around childbirth.

There are various social work specializations, but the larger categories include child, family, and school social workers, who provide social services and assistance to children and their families; medical and public health social workers who provide support for people with illnesses, such as Alzheimer's disease, cancer, or AIDS; mental health and substance abuse social workers who deal with people who struggle with psychological issues; and social workers who deal with the intricacies of social policy and planning.

CHALLENGES AND OPPORTUNITIES OF SOCIAL WORK

We all aspire to live in a society that is healthy, tolerant, safe, inclusive and fair. Such a society needs a vibrant economy and excellent public services. Social work services have an essential contribution to make in achieving that goal. Their role has three main dimensions:

* supporting the most vulnerable and excluded members of our society to live fulfilling lives, working in partnership with individuals, families and communities and with other public, voluntary and private services;
* protecting individuals, families and communities at risk of harm from themselves or others through the use of statutory powers, then working to reduce and minimise that risk through helping people to change their behaviours;
* working with others to close the opportunity gap between the richest and poorest in our society, through helping individuals and families

to take control of their lives, and develop hope and aspirations for the future.

Effective social work services promote independence and resilience, enabling some of our most vulnerable, excluded and even dangerous people to play an active part in society, through achieving change. In this way, they contribute not only towards the goal of delivering excellent public services, but towards developing the economy, helping people to become self reliant once more.

INTEGRATION AND SOCIAL WORK SERVICES

All public services are increasingly working towards integrated service delivery. The theme of services designed and delivered around the needs of individuals, families and communities is at the heart of much policy and has already had significant impact on social work services. Developments in community care are perhaps the furthest advanced, through the Joint Future initiative and single shared assessment. Similar approaches underpin community schools and proposals for integrated assessments for children. Joint approaches to planning through children's service plans, community care plans and ultimately community plans provide a vehicle for joint development. Structural approaches such as the creation of community justice authorities and community health partnerships take that integration further into new partnership arrangements. All of this has a direct impact on the structure, planning and delivery of social work services. A significant challenge for social work services will lie in maintaining and developing the distinctive contribution of social service workers, while promoting joined up approaches to service design and delivery. The three branches of social work will face distinctive challenges, reflecting the changing needs and expectations of society:

Children and Families

Family breakdown, fluid and unstable relationships, more children suffering chaotic lifestyles due to substance abusing parents, or from adults who have never learned to parent, and the growing numbers of disabled children living longer have all combined to create a growing number of children with a level and complexity of need barely imagined in 1968. Our understanding and recognition of child abuse, in particular child sexual abuse, mental health problems amongst children and the impact of domestic abuse on child development have grown significantly. Several high profile tragedies, service failures and enquiries have given child protection a very high political and media profile. This has led to a lack of acceptance of risk in the media and the public eye and a scapegoating of social work services when things go wrong. However there is also a growing understanding among partners that social workers alone cannot protect all children in need and that the universal services of health and education must take a greater role in delivering targeted services. Social workers are increasingly working in integrated early years services alongside health visitors and child care workers and in schools alongside teachers and school nurses, a trend which seems certain to continue. While residential child care remains the option of choice for a few children, many more are accommodated in residential provision due to a shortage of foster placements. The increasing complexity of problems faced by these children emphasises the need for highly skilled practitioners in the residential child care setting.

Youth crime and anti-social behaviour are high on the political agenda. Society is facing the consequences of rearing a group of disaffected, alienated young people with little stake in our society, though there is also some evidence of a growing intolerance of young people. Addressing this is the responsibility of all public services, not just social work services, which often deal with the consequences of alienation. Social workers will be major players in an integrated approach to youth justice. Their skills in engaging, persisting with and managing change in the most resistant individuals will be vital within partnership approaches to one of our society's most stark challenges.

The pressure on children and family social workers has been made worse, in recent years, by shortages of social workers able and willing to work in this demanding and highly pressurised environment. While the situation is improving, it has left a legacy in some authorities of unallocated cases, high thresholds before a service can be provided, inexperienced frontline staff, managers and a culture of crisis intervention.

Community Care

Since the closure of large institutions in the 1990s, social workers, occupational therapists and social care staff have made huge positive contributions to the development of better adult care services, increasingly in partnership with NHS colleagues. This has meant:

* new assessment processes;
* new ways of delivering rapidly expanding services;
* opening up a diverse provider market; and
* managing huge budgetary transfers.

Many people who were receiving institutional care have exceeded society's expectations of their ability to live fulfiled lives in the community. Innovative services have evolved to enable them do so. The increasing aspirations of previously marginalised groups and the subsequent increased demand for services are a consequence of past successes.

Adults with learning disabilities increasingly expect to live ordinary lives in the community. Services have to be open to developing their aspirations, but must protect those who may be vulnerable from bullying and challenging behaviour. This goes beyond individual casework and will challenge society's attitudes, especially around sexual relationships and parenting. For people with mental health problems, many of the themes are similar. The incidence of suicide and self harm in Scotland challenges how we make services available. Services will be shaped by the increasing emphasis on rights and responsibilities of people experiencing mental illness, and the needs of their carers and families. An integrated approach to improved mental health will require further development of partnership working with health, housing and other services.

People with physical and sensory impairment will continue to challenge services to personalise care. Good work is already being done using technology to increase capacity for independent living, driven by a rights based approach to services.

Similar challenges face the delivery of services for older people, in helping them maintain healthy, fulfiled and independent lives in the face of greater prevalence of long term conditions. This means that services must work together to promote self care and enable early responses to problems, reducing isolation and finding truly personalised solutions to people's needs.

People with substance misuse problems have needs which cross service boundaries, further challenging services to develop personalised approaches that make effective use of specialist expertise while meeting people's health needs, addressing the impact of the substance misuse on children and families and protecting the community from any associated offending behaviour. A better informed public will have growing expectations of public services and will be more knowledgeable and confident in producing their own care solutions with appropriate support.

Developments such as direct payments will place more emphasis on self assessment and self-care strategies to support independent living, and on the role of unpaid carers and local communities as providers of care. The introduction of an unregulated workforce, in the shape of workers employed directly by people who use services, needs to be fully understood and managed.

All this will only be achieved by stronger partnership between social work services, health, housing and others, within a mixed economy of public voluntary and private sector provision in which all parties have a stake.

Criminal Justice

Demand on criminal justice social work services will continue to grow, driven by the need to reduce serious and violent crime, the desire from communities for decisive action on anti-social behaviour and the need for a wider and more robust range of non-custodial options to tackle an ever increasing prison population. Increased awareness of sexual abuse will challenge services to balance public protection with the rights and needs of their clients, requiring closer partnership between police and criminal justice social workers to manage high risk offenders.

SOCIAL WORK NOW AND IN THE FUTURE

The social service workforce delivers essential services day in and day out to some of our most vulnerable people. Theirs is life changing work. We have found countless examples of social work services which transform people's lives and protect them and their communities from harm. We have found dedicated staff going way beyond what is expected of them, sacrificing their own time and personal energies to support people in need. Most of this work goes unnoticed and unremarked and it is important that we celebrate the skillful and innovative work seen in current practice.

It is also important that we are clear about the size of the challenge to develop social work services which meet growing and increasingly complex demands. This chapter looks at the strengths we can draw on in meeting that challenge and the areas where we want to see change. This is based on evidence (summarised in Annex A) from a wide and diverse range of sources including from:

* practitioners and leaders in the public, voluntary and private sectors;
* people who use services and their carers;
* partner professions and agencies;
* observing practice; and
* research.

We found many strengths in social work services, which need to be kept and developed as a basis for delivering future services. In particular we found:

* many examples of good practice across the whole range of social work services, often not acknowledged or celebrated;
* a social work profession with a strong and highly relevant knowledge, skills and value base, and a passion for social justice, which is attracting an increasing number and diversity of new recruits;
* professionals working in the most challenging of circumstances, balancing conflicting needs and views, juggling resources and making

the finest of professional judgements about risk to both individuals and society;

* a strong drive for change through the creation of an honours degree in social work, the Scottish Social Services Council and the Scottish Institute for Excellence in Social Work Education, working together to raise the professionalism of the workforce and improve the competence and quality of social workers' practice;
* services which deal skillfully with complex problems, helping people to find a way through their difficulties;
* a real commitment to developing joint services and a recognition among partners of the distinctive contribution that social service workers make to multi-agency working;
* a mixed market of services across the public, voluntary and private sectors each with the scope to learn from each other;
* an appetite for performance improvement and revitalised inspection; and
* that most people who use services were generally satisfied with the service they received.

MANAGEMENT TRAINING AND DEVELOPMENT IN INDIA

The liberalization and globalization process, which has come over the Indian economy in the last sixteen years, has given birth to a fair number of fly-by-night small enterprises which hope to gain legitimacy and foster rapid growth by setting up a Human Resource Department. This book, based on the authors' own experience, portrays, in an unrestrained manner, the fate that could easily overtake the unfortunate H.R. professional.

As Sadri, Ray and Hegde had argued, winds of change have swept the economy and the polity of India. The country has moved towards a free market economy. Old rules of the game have been discarded and new rules have been made. Old citadels of power have made way for new baron to take charge. The Nehruvian concept of a socialistic pattern of society,(weather that meant), has been replaced by an IMF influenced liberalization policy. Indian economic thought has undergone radical charges. It is in this climate of change that we seek to look at Organizational Restructuring, Human Resource Development, Training and Development and what they mean for the employees. The article champions no cause nor does it wave any manifesto. All that is attempted is a reexamination of a particular aspect of the Human Resource Professional in a changed environment over which he has little control.

DEVELOPMENT OF THE PERSONNEL AND H.R. FUNCTION IN INDIA

The 1950s in India saw the birth of the Personnel Function when outdated technocrats and retired army officials were appointed by industry to look after workers' interest. The 1960s saw a stress on the welfare attitude in the personnel profession. Then Labour Officers became interested in looking after canteens and latrines, meals and uniforms.

The 1970s witnessed the introduction of the legal angle so that the adjudication of disputes became important and personnel officers went about obtaining law degrees to horn their legal skills for use on the shop floor. The 1980s saw an enlargement of the legal aspect, so that collective bargaining became the most important skill of the Personnel Officer. Bipartite negotiations were complemented by union militancy during the first half of the decade and by management militancy during the second half. The 1990s saw the era of participative management on the one hand and the slow withering away of unions on the others. To use the terminology of the legendary C Wright Mills, union leaders became managers of discontent, while the bourgeoisie infiltrated the ranks of the managers.

Economic Liberalization

Ray and Sadri have already pointed out that since India began to liberate its economy and its market in 1991, a certain amount of uncertainty has crept into some segments of the economy. One prominent segment is the so-called "public sector" which really is a form of state capitalism.

It was "protected" by a variety of concessions, all in the private sector were often taken over by this sector, to preserve employment at all costs instead of allowing them to liquidate their assets. In the case of basic and key industries as listed under the Industrial Policy Resolution of 1956,rigid bureaucratic controls and a lack of accountability generally kept *per capita* productivity miserably low. There was, however, a limit beyond which the economy could not afford to main these non-profit making centers on the dole, especially when the country itself was passing through an economic crisis as in the early 1990s.

When India finally had to seek a financial bail-out from the World Bank and the IMF, certain conditionalities were imposed on the Government. One of these was to liberate the markets, to encourage private industry, and open the economy of the world. Now when you have tied up the hands and feet of a man for 45 years and you suddenly untie them and tell him to compete with Carl Lewis, will he succeed? However, the Indian Government and the news media, after years of protected state capitalism, wanted liberalization to succeed over-night.

Side Effects of Liberalization on H.R.D.

Scholars and managements were busy eulogizing the virtues of "liberalization" and the economic and social benefits that would flow from a free market economy. However, in their zeal to pursue that elusive rainbow, they forgot the importance of the human resource. They, ideologues tended to be polarized. After all, an extreme form of capitalism produced a Hitler just as an extreme form of socialism gave birth to a Stalin, and under both conditions, the exploitation of the human resource was equally horrendous. This paper tries to show that the human resource can neither be taken for granted nor can its importance be ignored without society having to pay enormous costs in the future.

Firms, which had previously been hankering for licenses to produce for a captive market suddenly, changed their focus and began seeking greater efficiency and productivity at all levels. Gaining the competitive edge become the most important concern for all and sundry. Some of the bigger and more reputed firms retained their culture, ethics and ethos while several of smaller once began to cut corners. This paper, based on real life experiences, is concerned only with the latter, i.e. with the small firm, which is a relatively new player in the competitive market and which, is desperately trying to become a multinational overnight. No sideswipes are intended and no person or organization is being alluded to. Owners of such firms faced for the first time with dynamic market competition from local and international products did not wish to cut their fixed costs. So they began to cut their variable costs. When this was not enough to produce the profits they desired, they began to treat human resources as a variable cost to be cut down.

Let us here briefly recall the development of Human Resource Management (HRM) in welfare officers required by the Factory's Act, we have traveled a long way into the present day scenario when most organizations, if not of need, at least for fashion, employ someone exclusively to adorn the position of Human Resource Manager. It has taken a considerable period for employers really to realize the need for human resource managers in India; and just as we were in the process of implementing this realization, the markets began to be liberated and we are now faced with another turbulent period. Employers, faced with such tremendous changes in the environment had once again to grapple with the question of human relations in industry. How important is the human being? Is it necessary to reorder priorities and to see weather the earlier realization of the importance of human resource management needs modification. HRM once again occupied the limelight. One important point to bear in the mind is that in India the advantages reaped by industries in terms of low labour costs are overshadowed by even lower labour productivity and other factors such as high energy costs, absurdly high interest rates of borrowed capital.

This is the perhaps the best manifestation of executives having to take corporate decisions in a hostile external environment. One clear conclusion, however, is that HRD initiatives have perforce to be focused on building a vibrant corporate culture on the one hand and increasing the per capita productivity of the employees on the other. With this broad agenda, the HRD executive has his work cut out for him for another few years at least.

The hostile environment we have mentioned does little to attract to foreign capital which is vital for growth in a market economy. Even Indian investors are today shying away from the stock markets. Now if the circulation of money in the economy falls, employers cannot make profits. How then can industry flourished?

Yet, when the axe has to fall and blame is to be attributed the HRD executive makes an ideal " fall guy". Therefore, he tries to build a halo around his specialty by coining new terms and phrases like old wine in new bottles. Nevertheless one of the most mouthed buzz words in Indian industrial centers today is Human Resource Development (HRD), which rivals even "Business I Process Reengineering (BPR)".

Such is the popularity of HRD that it is not surprising to find an ex-receptionist masquerading as a consultant in human resources administration or an ex-salesman gifted with the gap elevating himself to the position of a Management Trainer. Part time and distance learning courses give them a sort of legitimacy to stake their claim and easy to read manuals enable them to show their expertise to a limited audience. The limited audience which is addressed, has a three-fold advantage from the consultant's point of view,. First of all, their knowledge of the subject is so limited that they can easily be taken in by training and are afraid that a bad remark from the trainer could affect their careers. Thirdly, they are generally so awestruck by the trainer that they accept to be fobbed off with a reporter or a clever joke when they have asked a serious question.

These disadvantages of the trainees are compounded by the sketchy knowledge on the part of several corporate managers regarding what exactly is HRD. This sometimes leads senior executives who are out of date and out of touch with their mother discipline to concentrate on HRD, which in their limited wisdom, they consider a soft option.

BUSINESS PROCESS REENGINEERING (BPR)

To argue that globalization without completing the liberalization process is a fatal economic flaw that is not understood by many corporate managers. They all wish to globalize their operations even though such things as the primitive 'octroi' duty remain part of the present economy five years after the reforms began! The Spate of reforms ushered in over the past years, as we know, have forced top management to sit up and think. Small time entrepreneurs trading on the margin and speculating with venture capital have suddenly found out they can become very rich. The urge to grow richer takes them to consultants who often are professional dilettantes of the same ilk as the H.R. "specialists" alluded to. These "consultants" have a sure for all ills, a panacea for all organizational needs – B.P.R. or business process reengineering.

Once top management has been made to swallow the bait, the organization is committed to BPR hook line and sinker. Top management then insists on a sea change in attitudes overnight and expects employees to think in terms of teams, goals, processes and HRD is expected to perform the magic to make all this happen. Many a time top management has no clue what HRD is all about and yet they seek HRD interventions.

The CEOs and MDs if such firms often say "the only thing constant in this company is change", and what is most damaging to the organization is that all change is seen as synonymous with progress. The CEO quite often finds himself on the horns of a dilemma when he wants the organization to change with the times but is unable to do so himself. The sweep back can be catastrophic the CEO pushes the organization forward, but his own unwillingness to let go of power creates contradictions at all levels in the organization.

Empowerment and other Contradictions

The first instance of contradiction is seen when top management speaks of empowerment but is unwilling to let senior executives decided without a nod from the CEO. This means that, whereas, the organization is supposed to be flexible to keep in tune with market realities, its executives are bound by old authoritarian structures. Thus, opportunities are lost, decisions are made based on individual preferences, power games are played and sycophancy thrives. Mediocrity is safe, as it poses no threat, to those at the top.

The second contradiction is to be found when CEOs speak of trust, faith and delegation but at the same time have a very powerful "internal audit" department whose ambit of concern spreads far beyond checking accounts. Is it a surprise then that good chartered accounts shy away from internal audit? Only those who cannot make it elsewhere or have a nasty streak in their character often opt for this, if only to increase their nuisance value. When internal audit gains precedence in this fashion all HRD initiatives will go out of the windows, because the small employer understand the financial bottom line better than the needs of human beings. Since every activity is now reduced to monetary terms, Internal Audit is concerned with everything. In short it is the eyes and ears of top management, its men are universally abhorred and kept at arms length by every professional: accountant, technologist human resources expert, systems analyst and others in the organization. Some organizations have even moved to a higher level of sophistication in management by remote control and call this department Corporate Governance.

The Third Example of contradiction can be found when CEOs speak of Transparency and Openness while almost every sensitive aspect of management remains a closely guarded secret. In short, transparency at the lower level is a must, but transparency at the top is a term reserved exclusively for business meetings and managers conferences. It is quite common, for instance, to find a company claiming to have one foot in the 21st century when there is trusted blood relative of top management placed in each department to give regular and personalized feedback on the others.

The list of contradictions, though quite long, need not be further lengthened. Suffice it is to say that these CEOs and those close to them, hardly if ever," walk their talk". What invariably happens is that small pocket of power emerge with their own informal rules and their special bonds of loyalty where the mediocrities with one another for greater influence.

Effect on the HR Professional

In their bid to appear progressive, these companies which are trying to move from the status of private businesses to corporate houses often recruit top class H.R. professionals. However, the fact is that the transition from private ownership and control to a modern organization is prevented from happening by the sheer weight of external contradictions which takes time to sink into the consciousness of these recently hired professionals. By the time this happens, the damage is done and if the professional is sloe to act, the damage is compounded.

At this point yet another contradiction emerges: that between what the HR professional wishes to achieve and is able to achieve. First of all, the professional frame logical polices which will hopefully improve structures and functions. However, all policies need approval from the CEO and this can take up to six months on even minor issues. The HR Professional then slowly realizes that, unless old rules are discarded, any reorganization achieved will be no more effective than dusting the furniture in Pompeii.

Perhaps the greatest handicap the HR Professional faces is the mediocrity he has inherited which with its power broking capacity will now allow its incompetence to be made public. Hence sabotage begins at the door of the H.R. Division. The pulse rate of the H.R. H.R. Professional increases when he finds that every provision of labor legislation has been flouted with impunity. Premises are not registered either under the Factories Act or under the Shops and Establishments Acts; Provident Fund returns have not been submitted and it is anybody's guess where the funds have gone; graduate's is being paid below the statutory minimum rate; the provisions of the Employee State Insurance Scheme are quietly side stepped. And, when inspectors do call in, the management unite to decry what is emphatically called the "inspector raj". More freedom is demanded without specifying freedom to do what"?

The HR person's heightened pulse rate leads to a higher blood pressure when attempts to pay for performance are subjected to top management's perception of performance at the individual level. In comes the internal audit man or the "blood relative" to decide who has performed and who has not. Appraisal forms are filled, data may or may not be analyzed, but increments operate on their own logic.

The unkindest cut of all comes when despite weak fundamentals the CEO announces that the Company must grow from strength to strength, and in process more manpower needed. To give the illusion of growth, revenue earned, from the sale of equity shares is included in the turnover. Much to the chagrin of the HR professional who advises caution, he is told that either he is not effective or possesses no business sense. All this is fine until the market no longer swallows the pill and the product does not sell. Suddenly the bubble bursts and the reality of the market overtakes the mirage of the visionary CEO. The HR professional is then the fall guy who has to go ahead with reduction in manpower, agree to increments not necessarily based on the performance and justify the unjustifiable. What happen to HR professional? Does he make a transition from being an icon worshipper to an iconoclast? Does he resign to himself to his fate and become "his master's voice"? Or does he beat a speedy exit in the hope of flourishing elsewhere? And should he quite, what prevents his past employer from floating nefarious whispers against him in the employment market? Whatever he finally does, the HR professional will end up being a better judge of corporate CEOs who wish to develop too fast for their own good. What of the rest of the employees?

Well they become unwitting canon fodder, but what of the mediocrities? They are safe since they are willing to cajole, please, adept and carry on. But the greatest damage is inflicted on the national economy. Since it is the economy which permitted these companies to get rich quickly, it is the economy that must pay for their misdemeanors.

The liberalization theorists must not lose sight of the fact that when they sow the wind, they will also reap the whirlwind. So much for the political economist on the either end of the ideological polarity to munch over. As for the HR professional, he is caught in a cleft stick. He can not take sides, is expected thoroughly ethical and prepared to bear his own cross. He is that change agent whom the CEO afraid to empower. He fills like a fish out of water when surrounded by the mediocrities in whose company the CEO is quit comfortable.

Insecure trade unionists resist change and employ every trick in the book to retain status quo. The insurance first line supervisor aids this resistance. In organizations where trade union tendency is thankfully not present, the existence of mediocrity is sometimes compounded by the false consciousness of the management and the aristocracy of labor leaders from the outside the organization who are more interested in money, symbol of status, designations etc. than achieving the real progress. Trade unions then become super become superfluous, not so much because the growing consumerisms of the work force make them think unions can be dispensed with. Managements become more paternalistic and only mouth platitudes of progress from time to time. And lack the means of communicating ideas of the true progress. Organizational change, must of necessity, is form the top ten downward. But will the top echelons of management accept change in regard to Human Resource Development. The CEO likely to fall back on the fact that although wages are low, labour productivity is also low, interest rates on capital are high and energy costs are exorbitant. He will be more likely to watch the bottom line than promote human relations.

Bibliography

Ainapur, L. S.: *Dynamics of Caste Relations in Rural India*, Jaipur, Rawat Publications, 1986.

Ali, Asghar : *Mandal Commission Controversy*, Delhi, Ajanta Publications, 1991.

Allen, John C. and Dillman, Don : *Against All Odds: Rural Community in the Information Age*, Boulder: Westview Press, 1994.

Ashworth, Gregory J.: *Marketing Tourism Places*, London, Routledge, 1990.

Basanti Das: *Governmental Programmes for Rural Development*, Discovery, Delhi, 2007.

Benjamin, Joseph: *Scheduled Castes in Indian Politics and Society*, New Delhi, Ess Ess Publications, 1989.

Bhai, N.: *Harijan Women in Panchayati Raj*, Delhi, BR Publishing 1986

Cootner, Paul : *The Random Character of Stock Market Prices in Rural India*, MIT Press, 1964.

Coppack, P. M.: *Entrepreneurial and Sustainable Rural Communities*, Mt. Allison University Press 1990.

Damon, William : *Social and Personality Development, Infancy through Adolescence*, New York: Norton, 1983.

Daniel, R.: *Recent Advances in Tourism Marketing Research*, New York, Hawthorne Press, 1996.

Derrett, J. Duncan. *Religion, Law, and the State in India*. London: Faber, 1968.

Desmond, Glenn, and John A, Marcell, *Handbook of Small Business Valuation Formulas and Rules of Thumb*, Valuation Press, 1993.

Diwan, Jagmohan : *Fundamentals of Rural Sociology*, Cyber Tech Pub, Delhi, 2009.

Doshi, S.L. & Jain, P.C. : *Rural Sociology*, Rawat, Delhi, 1999.

Durkheim, Emile: *The Division of Labor in Society*, New York: Free Press, 1964 .

Erikson, E. H.: *Panchayati Raj and Society*, New York, W. W. Norton, 1993.

Farquhar, J.N.: *Modern Religious Movements in India*, Munshiram, New Delhi, 1967.

Gawthrop, L.C. : *Public Sector Management Systems and Ethics*, Bloomington, Indiana University Press, 1984.

Gies, Frances : *Women in the Middle Ages*. New Jersey: Barnes and Noble, 1978.

Gowdy, J., *Coevolution Economics: The Economy, Society and the Environment*, Kluwer, Boston, 1994.

Granger, Clive W. J.: *Rural Marketing in Economics, Specification and Evaluation*, London, Cambridge University Press, 1999.

Guala, F.: *The Methodology of Rural Economics,* New York, Cambridge University Press, 2005.

Guarti, Luigi, *The Rural Management of Marketing,* Blackwell Publishing, 1994.

Gurukal, Rajan: *The Formation of Caste Society in Kerala,* Rawat Publications, New Delhi, 1994.

Harman, William: *The Sacred Marriage of a Hindu Goddess,* Indiana University Press, Bloomington, 1989.

Harrigan, K. R.: *Strategies for Declining Businesses.* Lexington, MA: Heath, 1980

Jain, D.: *Women's Quest for Power in Panchayati Raj,* Sahibabad, Vikas, 1980.

James Makens: *Marketing for Hospitality and Tourism*: New Jersey, Prentice-Hall, 1998.

Jeff Hass: *Economic Sociology: An Introduction,* Manohar, Delhi, 2008.

John, D. Kasarda: *Contemporary Urban Ecology,* New York: Macmillan, 1977.

Karve, Irawati: *Hindu Society: An Interpretation,* Poona, Sangam Press, 1961.

Louis, Prakash : *The Political Sociology of Dalit Assertion,* Gyan, Delhi, 2003.

MacDonald, S.: *Evolutionary Perspectives on Human Development,* Thousand Oaks, CA: Sage Publications, 2004.

Madan Mohan: *Impact of Integrated Rural Development Programmes,* Omega Pub, Delhi, 2007.

Marris, R. L., and Wood, A.: *The Rural Marketing Economy,* London: Macmillan, 1971.

Mathur, B.L.: *Rural Management,* RBSA Pub, Delhi, 2008.

Meyer, M. W.: *Theory of Organizational Structure in Rural Marketing,* Indianapolis, 1977.

Mueller, R.: *Women's Work in Third World Agriculture,* Geneva, ILO, 1985.

Rabi N Misra: *Banks for Rural Development,* Sonali Pub, Delhi, 2007.

Ramesh, G S : *Philosophy of Indian Sociology,* Global Vision, Delhi, 2005.

Robert, T.: *Practical Public Management,* New York, Marcel Dekker, 1995.

Sharma, Kususm: *Ambedkar and Indian Constitution,* New Delhi, Ashish, 1992.

Silverberg, James: *Social Mobility in the Caste System in India,* Mouton Publishers, The Hague, 1968.

Singh, Balveer : *Medical Sociology of Tribal Community,* Himanshu, Delhi, 1992.

Srinivas, M. N.: *Religion and Society Among the Coorga of South India,* Oxford University Press, London, 1952.

Tietenberg, Tom, *Environmental Economics and Policy of Rural Marketing,* New York, HarperCollins, 1994

Towse, Ruth ed.: *A Handbook of Rural Economics,* Edward Elgar, 2003.

Vijay Kumar: *Rural Development Participatory Human Management,* Anmol, 2009.

Weibull, J. W.: *Advances in Understanding Strategic Behaviour,* New York, Palgrave, 2004.

Woodward, J.: *The Oxford Handbook of Philosophy of Economics,* New York, Oxford University Press, 2009.

Yatindra Singh Sisodia: *Rural Development : Macro-Micro Realities,* Rawat, Delhi, 2007.

Index